ÍNDICE

Nota del autor

This material contains enough theoretical development and proposed exercises with solutions, in order to promote our students' autonomy. Moreover, for almost each introduced concept, some exercises are solved step by step.

To prepare this material, I have used the following bibliography:

Bibliography

- Fundamentals of Mathematics (Authors: J. Van Dyke, J. Rogers, H. Adams; Ed.: BROOKS/COLE, CENGAGE Learning),
- Spanish text books:
  - Matemáticas 2º ESO (Autores: J. Cólera y otros; Ed.: Anaya).
  - Matemáticas 2º ESO (Autores: A. M. Gaztelu y A. González; Ed.: Santillana).
- Notes and work sheets (in English and Spanish): Alfonso González, José A. Jiménez Nieto, Miguel Ángel Hernández Lorenzo, Isabel Torres (IES Ruradia), IES Don Bosco.
- English Dictionary (Oxford Edition).

Esta edición del libro tiene muchísimos menos errores tipográficos, de idioma y de cálculo gracias al profesor **D. Mariano Salido Esparza**. Mi más sincero agradecimiento. Profesores con esta dedicación hacen grande esta profesión.

This materials have been reviewed in order to fulfil new Spanish **LOMCE** regulations.

Para el profesor: Evaluación por estándares: He preparado un cuadro en el que relaciono cada estándar evaluable de la LOMCE con los ejercicios de este libro que lo trabajan, y que puede ser muy útil a la hora de diseñar las pruebas de evaluación. Si estás interesado en que te lo envíe, ponte en contacto conmigo en el mail de más abajo.

En mi web (www.cuadernodepitagoras.com) he subido las presentaciones que uso con mis alumnos en clase. Si en algo crees que te puedo ser de utilidad, si encuentras erratas, o si deseas hacerme alguna sugerencia, te agradeceré que te pongas en contacto conmigo en el mail javiersanchezpi@gmail.com.

Soy doctor en Química y licenciado en Ciencias Ambientales y soy profesor de matemáticas en un instituto en Murcia (España).

Espero te sea de utilidad y que contactes conmigo para lo que precises.

Un saludo.   Javier Sánchez Pina.

Unit 1.- Natural numbers

1. Remember how to read numbers

First of all, we are going to remember how numbers are read in English. If necessary, look at your notes of last course. Anyway, we are going to remember now. At the beginning of some units, we will also remember whatever we need to know about reading.

Ordinals and cardinals

Remember natural numbers are classified, depending of their utility, into:

– Counting: "There are three apples on the table" → CARDINAL NUMBERS

– Ordering: "Barcelone is the second largest city in our country"→ORDINAL NUMBERS

Cardinal Numbers	Ordinal Numbers	Cardinal Numbers	Ordinal Numbers
1. one	1st. First	20. twenty	20th. Twentieth
2. two	2nd. Second	21. twenty-one	21st. Twenty-first
3. three	3rd. Third	22. twenty-two	22nd. Twenty-second
4. four	4th. Fourth	23. twenty-three	23rd. Twenty-third
5. five	5th. Fifth	24. twenty-four	24th. Twenty-fourth
6. six	6th. Sixth	30. thirty	30th. Thirtieth
7. seven	7th. Seventh	40. forty	40th. Fortieth
8. eight	8th. Eighth	50. fifty	50th. Fiftieth
9. nine	9th. Ninth	60. sixty	60th. Sixtieth
10. ten	10th. Tenth	70. seventy	70th. Seventieth
11. eleven	11th. Eleventh	80. eighty	80th. Eigthtieth
12. twelve	12th. Twelfth	90. ninety	90th. Ninetieth
13. thirteen	13th. Thirteenth	100. a/one hundred	100th. Hundreth
14. fourteen	14th. Fourteenth	101. a/one hundred	101st. Hundred and first
15. fifteen	15th. Fifteenth	200. two hundred	200th. Two hundreth
16. sixteen	16th. Sixteenth	1 000. a/one thousand	1 000th. One thousandth
17. seventeen	17th. Seventeenth	10 000. ten thousand	10 000th. Ten thousandth
18. eighteen	18th. Eighteenth	100 000. a/one hundred thousand	100 000. One hundred thousandth
19. nineteen	19th. Nineteenth	1 000 000. a/one million	1 000 000 One millionth

World's cultures

Beyond a million, the names of the numbers differ depending where you live.
The places are grouped by thousands in America and France, by the millions in Great Britain, Germany and Spain.

Name	American-French	English-German-Spanish
million	1,000,000	1,000,000
billion	1,000,000,000 (a thousand millions)	1,000,000,000,000 (a million millions)
trillion	1 with 12 zeros	1 with 18 zeros
quadrillion	1 with 15 zeros	1 with 24 zeros

AND

AND is used before the last to figures (tens and units) of a number.

> 103: a (or one) hundred **and** three
>
> 325: three hundred **and** twenty-five.
>
> 2315: two thousand three hundred **and** fifteen

A or ONE

The words hundred, thousand and million can be used in the singular form with "a" or "one", but not alone.

"A" is more common in an informal style; "one" is used when we were speaking more precisely.

> I want to live for **a** hundred years
>
> The journey took exactly **one** hundred light years
>
> I have **a** thousand euros

"A" is also common in an informal style with measurement-words.

> A kilo of oranges costs **a** pound
>
> Mix **one** liter of milk with **one** kilo of flour ...

SINGULAR OR PLURAL?

Number are usually written in singular.

> Three hundred euros
>
> Several thousand light years

The plural is only used with dozen, hundred, thousand, million, billion, if they are not modified by another number or expression (for example a few/several).

> Hundreds of pounds
>
> Thousands of light years

PHONE NUMBERS

Each digit is said separately (25 → two five)

The figure '0' is called oh (304 → three **oh** four)

Pause after groups of 3 or 4 figures

>234 7809 → two three four, seven eight oh nine

>234 5778 → two three four, five double seven eight (British English) or two three four, five

>>seven seven eight (American English)

ZERO, NOUGHT, NIL, LOVE ...

The figure 0 is usually called **nought** in British English and **zero** in American English

In measurements, 0 is called zero:

>Water freezes at zero degrees Celsius

In team games, zero scores are usually called nil in British English, and zero in American English.
In tennis, the word love is used instead of zero (this is derived from French word "l'oeuf", because zero can be egg-shaped):

>Spain three Germany nil (zero)

>Nadal is winning forty-love

1 BILLION

In English, billion usually means a thousand million: 1 000 000 000
Realize that in Spanish billion means 1 000 000 000 000

PLACE VALUE

Every digit in a number represents a different value depending on its position.

For example: In 53, "5" represents fifty units

In 5234, "5" represents five thousand units

This is the place value table we need to write numbers, no matter how big they are:

hundred ten one	hundred ten one	hundred ten one	hundred ten one
Billion	milllion	thousand	(unit)

CALCULATIONS

ADDITION $2 + 4 = 6$	• Two **and/plus** four **is/are/equals** six • Two **added to** four **makes** six • **What's** two **and** four? **It's** six.
SUBTRACTION $8 - 5 = 3$	• Eight **minus** five **is/are/equals** three • Eight **take away** five is three • Five **from** eight **leaves/is** three
MULTIPLICATION $6 \cdot 5 = 30$	• Six **times** five **is/equals** thirty • **Six fives are** thirty • Six **multiplied by** five **is/makes** thirty (More formal way)
DIVISION $12:3=4$	• Twelve **divided by** three **is/are/equals** four • Three **into** twelve **goes** four times (for smaller calculations)
POWERS 6^5 (6 is the base and 5 is the index or exponent)	• Six **to the power** of five SPECIAL • Six to **the fifth power** POWERS • Six **raised to** fifth 5^2 : Five squared 4^3 : Four cubed
ROOTS $\sqrt{16} = 4$	• The **square root** of sixteen **is/equals** four.

1. Write the following numbers in words as in the example:

a) 3456   b) 90304   c) 765   d) 237   e) 98053

f) 134008   g) 45004   h) 150003

2. Write the following numbers in words as in the example:

a) 3528   b) 86 424   c) 987   d) 3270   e) 30001

f) 1487070   g) 320569   h) 20890300

3. Write the missing words. Then, write the answers in numbers and symbols:

Example: Ten plus three equals *thirteen 10+3=13*   .

a) Twelve minus six equals _____

b) Seven times one equals _____

c) Twenty-five divided by five equals _____

d) Eight plus four minus nine equals _____

4. Write the missing numbers. Then, write the answers in words.

Example: $3+8 =$ _11. Three plus eight equals eleven._

a) $3 \cdot$ ___ $= 30$ _____

b) ___ $-5 = 13$ _____

c) ___ $+2 = 4$ _____

d) ___ $:2 = 10$ _____

5. Write the missing symbols. Then, write the answers in words.

3___7___$4 = 14$ _____

9___2___$2 = 20$ _____

25___5___$4 = 9$ _____

Exercises

6. Calculate the following powers mentally and write them in words:

Example: $4^3 = 64$ *Four cubed equals sixty-four*

a) $5^4 =$ _____

b) $11^2 =$ _____

c) $2^5 =$ _____

d) $5^3 =$ _____

e) $10^3 =$ _____

f) $100^2 =$ _____

7. Calculate mentally and write in words as in the example:

Example: $\sqrt{16} = 4$ *The square root of sixteen is four*

a) $\sqrt{25} =$ _____

b) $\sqrt{121} =$ _____

c) $\sqrt{900} =$ _____

d) $\sqrt{1600} =$ _____

2. Divisibility relationship

A *multiple* of a whole number is the product of that number and a natural number. Also, a natural number is multiple of another if it is in its multiplication tables. *Multiple of* is synonymous of *divisible by*.

So, we can see it in two ways:

For example: For example:

- multiples of 2 are 2, 4, 6, 8, …   • 21 is a multiple of 7 because 7 x 3 = 21

- multiples of 3 are 3, 6, 9, 12, …   • 77 is a multiple of 7 because 7 x 11 = 77

- multiples of 4 are 4, 8, 12, 16, …   • 98 is a multiple of 7 because 7 x 14 = 98

A whole number that divides exactly into another whole number is called a **divisor** of that number.

For example 20 : 4 = 5. So, 4 is a divisor of 20, as it divides exactly into 20.

Notice: If a number can be expressed as a product of two whole numbers, then these whole numbers are *divisors* of that number.

Divisibility relationship

If number A is a **multiple** of number B, then B is a **divisor** of A.

Example 1: a) Is 60 a multiple of 20?   b) Is 20 a divisor of 60?

Solution: a) Yes, 60 is a multiple of 20, because 20·3 = 60.

b) Yes, because if we divide 60 by 20, quotient is 3, and it is an exact division.

Exercises

8. Write five multiples of 15.

9. Write five multiples of 14.

10. Write eight first multiples of 9.

11. Write nine first multiples of 12.

12. Write all the divisors of number 12. How many divisors does it have?

13. How many divisors does number 60 have?

14. Indicate how many divisors number 45 has.

15. Write all the divisors of 90.

Exercises

16. Write all the divisors of 75.

17. Write all the divisors of 18.

18. a) Is 40 a multiple of 10?   a') Is 10 a divisor of 40?

   b) 16 a multiple of 3?   b') Is 3 a divisor of 16?

19. Complete the following sentences:

   a) 24 is divisible by ...   b) 2 and 3 are divisors of ...

   c)   ... is a multiple of 5   d) Indicate 2 multiples of 11.

   e)   A divisor of 13 is: ...   f) …. Is a multiple of 37.

20. Explain in a clear way why 184 is a multiple of 23.

21. Is 17 a divisor of 255? Why?

22. Find out three numbers being multiples of 25.

23. Find out three divisors of 30.

24. a) Is 200 a divisor of 1 000?   Is 1 000 a multiple of 200?

   b) Is 15 a divisor of 70?   Is 70 a multiple of 20?

   c) Is 90 a multiple of 6?   d) Is 125 a divisor of 1000?

25. Write down:

 a) The multiples of 20 between 150 and 210.   b) One multiple of 12 between 190 and 200.

26. Write down:

   a) Write down all the multiples of 6 between 20 and 70

   b) Write down all the multiples of 7 between 30 and 80

   Write the three smallest multiples of 8 which are over 50

3. Tests of divisibility

One number is divisible by:

2 If the last digit is 0 or an even (0, 2, 4, 6, 8).

3 If the sum of the digits is divisible by 3.

5 If the last digit is 0 or 5.

11 If the sum of the digits in the even position minus the sum of the digits in the odd position is 0 or divisible by 11.

Exercises

27. Given following numbers, select:

| 66 | 71 | 90 | 103 | 105 |
| 156 | 220 | 315 | 421 | 708 |

a) Multiples of 2 →   b) Multiples of 3 →   c) Multiples of 5 →

28. Copy these numbers in your notebook, round with a circle multiples of 3 and cross multiples of 9. What do you observe?

$$33 \qquad 41 \qquad 54 \qquad 87 \qquad 108$$
$$112 \qquad 231 \qquad 341 \qquad 685$$

29. Without calculating, indicate which of the following numbers are divisible by 2, 3, 5, 10 and 11.

a) 29.304   Is divisible by:

b) 1.439.350   Is divisible by:

c) 34.530   Is divisible by:

d) 321   Is divisible by:

e) 507   Is divisible by:

30. Given following numbers, indicate which of them are multiples of:

239   300   675   570   800   495   888   6402   2088

a) 2:   b) 3:   c) 5:   d) 10:   e) 11:

31. Change every letter by the digit/s that make the number divisible by 3.

| A51 | 2B8 | 31C | 52D | 1E8 |

32. For each number, change every letter by the digit/s that make the number divisible, at the same time, by 2 and 3.

| 4 | *a* | | 3 | 2 | *a* | | 2 | 4 | *a* |

Divisibility tests for composite numbers

Given a composite number A $(A = a \cdot b \cdot c \cdot .... \cdot j)$, another number B will be divisible by A if it is divisible, separately, by $a, b, c, ...$ and j.

Example 2: Check why we can know 4020 is divisible   a) by 6,   b) 15 and   c) 30.

Solution:

a) Notice that $6 = 2 \cdot 3$,   4020 is divisible by 2 (last digit is an even number) and by 3 $(4 + 2 = 6)$.

b) Notice that $15 = 3 \cdot 5$, 4020 is divisible by 3 $(4 + 2 = 6)$ and by 5 (last digit is 0).

c) Notice that $30 = 2 \cdot 3 \cdot 5$,   4020 is divisible by 2, 3 and 5, so it is also divisible by 30.

33. Is 35,766 divisible by 6?

34. Is 11,370 divisible by 15? Write a divisibility test for 15.

35. Is 11,370 divisible by 30? Write a divisibility test for 30.

4. Prime and composite numbers

A *prime number,* or a *prime* is a whole number greater than 1 with exactly two factors (divisors). The two factors are the number 1 and the number itself.

A *composite number,* or a *composite,* is a whole number greater than 1 with more than two factors (divisors).

What about **0** and **1**?   0 and 1 are neither prime nor composite.

The sieve of Eratosthenes

A Greek mathematician, Eratosthenes (276 - 195 BC), discovered the Sieve which is known as the Sieve of Eratosthenes. It is a method to get prime numbers.

He got a table of whole numbers, e.g. from 1 to 100 and crossed out the number 1. After that, he crossed out number 2 and all multiples of 2. Later, 3 and its multiples, and he went on in this way.

Example 3: Copy the following table in your notebook and build the Eratosthenes' sieve.

1	2	3	4	5	6	7	8	9	10
11	12	13	14	15	16	17	18	19	20
21	22	23	24	25	26	27	28	29	30
31	32	33	34	35	36	37	38	39	40
41	42	43	44	45	46	47	48	49	50
51	52	53	54	55	56	57	58	59	60
61	62	63	64	65	66	67	68	69	70
71	72	73	74	75	76	77	78	79	80
81	82	83	84	85	86	87	88	89	90
91	92	93	94	95	96	97	98	99	100

5. Prime factorization

The *prime factorization* of a counting number is the indicated product of prime numbers.

To find the prime factorization of a counting number using repeated division

1. Divide the counting number and each following quotient by a prime number until the quotient
   is 1. Begin with 2 and divide until the quotient is odd, then divide by 3. Divide by 3 until the
   quotient is not a multiple of 3. Continue with 5, 7, and so on, testing the prime numbers in
   order.

2. Write the indicated product of all the divisors.

Example 4: find the prime factorization of the number 180:

Solution:

180	2
90	2
45	3
15	3
5	5
1	

So, prime factorization of 180 is: $2^2 \cdot 3^2 \cdot 5$.

Exercises

36. Find the prime factorization of the following numbers:

a) 270   b) 370   c) 630   d) 750   e) 1000   f) 1100

37. Find the prime factorization of the following numbers:

a) 48   b) 54   c) 90   d) 105   e) 120   f) 135   g) 180   h) 200

38. Work out the prime factorization of the following numbers:

a) 12   b) 18   c) 24   d) 36   e) 42   f) 360   g) 1350   h) 48

13

6. Least Common Multiple (LCM)

The *least common multiple (LCM)* of two or more whole numbers is the smallest natural number that is a multiple of each whole number.

The factorizations of this chapter are most often used to simplify fractions and find LCMs. LCMs are used to compare, add, and subtract fractions. In algebra, LCMs are useful in equation solving.

You can find the LCM of two or more numbers, using two methods:

Method 1 to calculate LCM of two or more numbers: Find the LCM of 6 and 9.

The multiples of 6 are: 6, 12, 18, 24, 30, 36, 42, 48, 54 ...

The multiples of 9 are: 9, 18, 27, 36, 45, 54, 63, 72, 81 ...

The common multiples of 6 and 9 are: 18, 36, 54, ...

So, LCM (6, 9) = 18.

Example 5: Try yourself: Find the Least Common Multiple (LCM) of the numbers 45 and 60.

Solution: The multiples of 45 are:

       The multiples of 60 are:

       The common multiples of 45 and 60 are:

       So the least common multiple of 45 and 60 is: LCM (45,60) =

Method 2: Using the Individual Prime-Factoring Method: Find the LCM of 45 and 60.

- Step 1: Find the prime factorization of each number in exponent form.

$$45 = 3^2 \cdot 5 \qquad 60 = 2^2 \cdot 3 \cdot 5$$

- Step 2: Find the product of the highest power of each prime factor.

$$LCM(45,60) = 2^2 \cdot 3^2 \cdot 5 = 4 \cdot 9 \cdot 5 = \boxed{180}.$$

39. Find Least Common Multiple of:

   a) 8, 12 and 9   b) 12, 10 and 9   c) 21, 6 and 14   d) 6, 33 and 22   e) 10, 15 and 12.

40. Find:

   a) LCM(8, 12)   b) LCM (48, 54)   c) LCM (90, 150)

   d) LCM (12, 15)   e) LCM (12, 30)   f) LCM (24, 60)

   g) LCM (40,90)   h) LCM (6, 10, 15)   i) LCM (8, 12, 18)

41. By a bus stop, two different lines are passing, A and B. Every 24 minutes a A bus passes, and every 36 minutes, one of line B. If one bus of each line have just coincided in this bus stop, when will they coincide again?

42. Pete has been told by the doctor to have a pill every 10 hours and a spoon of syrup every 24 hours. If he has just taken both of them together, when will both drug's administration coincide again?

43. Juan works in a bar, and when all the beer bottles of a box are empty, he puts the boxes forming a column. There are two types of beer, Mahou, whose boxes have height of 20 cm, and Coronita, whose boxes are 35 cm high. What is the shortest pair of columns he can form, having the same height? How many boxes of each type will he have used?

44. Jose earns 240 € per week. His brother earns 300 € per week, and his sister, 250 €. What is the least number of weeks each must work so that they earn the exact same amount of money?

45. A wire roll is more than 150 m long and less than 200 m. What is its exact length, knowing that it can be divided into pieces of 15 m and also in pieces of 18 m?

Exercises

15

The largest common factor of two or more numbers is called the **highest common factor** (HCF).

To write the highest common factor of two or more whole numbers using the Individual Prime-Factoring Method:
- Step 1: Find the prime factorization of each number in exponent form.
- Step 2: Find the product of the lowest power of **common** prime factors.

Example 6: Find the HCF of 120 and 100.

Solution: $120 = 2^3 \cdot 3 \cdot 5$ $100 = 2^2 \cdot 5^2$ $LCM(120,100) = 2^2 \cdot 5 = 4 \cdot 5 = 20$.

46. Calculate:

   a) HCF(36, 45)   b) HCF (48, 72)   c) HCF (105, 120)

   d) HCF (135, 180)   e) HCF (8, 12, 16)   f) HCF (45, 60, 105)

47. We want to divide a rectangular area of 120 m wide by 180 m long into square plots that are as large as possible. How much should measure the side of each plot?

48. In an athletic club have signed 18 boys and 24 girls. They want to make teams "unisex" as big as possible. How many members must have each team?

49. My family has just bought a rectangular field, where we want to build a camping. We want to divide the field into similar squares, being as large as possible. Dimensions of the field are 650 x 525 m. What will the dimensions of the squares be?

50. Marta has 12 red, 30 green and 42 yellow marbles and she wants to put them in boxes, as many as possible, all the boxes with the same amount of each colour and with no marbles remaining. How many boxes will she have? How many marbles of each colour are there in each box?

Exercises

52. We want to cut three wires with lengths of 275, 360 and 450 cm, into similar portions, being a long as possible. What will their lengths be?

53. We have covered the floor of a rectangular room (dimensions: 12.3 m x 9 m), by using squared tiles. You must know we did not have to cut any of them, and that we used the largest squared tiles in the shop. Calculate the side of the squared tiles.

LCM and HCF mixed problems

54. My baby has been playing with rectangular cards whose dimensions were 12 cm x 18 cm. With them, he has formed the smallest square. What is the side of the square? How many cards has he used?

55. We want to pack 175 coca-cola cans and 150 Fanta cans, in similar boxes, being as big as possible, and without mixing the products. How many cans will I put into each box? How many boxes do I need?

56. We want to bottle 100 liters of oil in identical bottles (their volume in liters is a whole number). Indicate all the possibilities we have got.

57. The three children of a family are visiting their parents with the following frequencies: the eldest comes every 15 days, the median one, every 10 days, and the youngest one, every 12 days. If they coincided on Christmas day, when will they coincide again?

Review exercises

1. True or false?

　a) 195 is a multiple of 13.

　b) 13 is a divisor 195.

　c) 745 is a multiple of 15.

　d) 18 is a divisor of 258.

　e) 123 is a divisor of 861.

Sol.: T, T, F, F, T.

2. Write:

　a) The five first multiples of 10.

　b) The five first multiples of 13.

　c) The five first multiples of 31.

3. Calculate:

　a) All the divisors of 18.

　b) All the divisors of 23.

　c) All the divisors of 32.

Sol.: a) 1, 2, 3, 6, 9, 18;　b) 1, 23;　c) 1, 2, 4, 8, 16, 32.

4. Write the first 5 multiples of 15 bigger than 1000.

5. Write all the divisors of 140.

Sol.: 1, 2, 4, 5, 7, 10, 14, 20, 28, 35, 70, 140.

6. True or false?

a) Sum of 2 multiples of 8 is a multiple of 8.

b) Difference of 2 multiples of 6 is a multiple of 6.

c) If a number is a multiple of 3 and 4, it will also be a multiple of 12.

d) If a number is a multiple of 20, it is also a multiple of divisors of 20.

Sol.: T, T, T, T.

7. Calculate the value/s of a so that number 71a :

a) Is a multiple of 2   b) Is a multiple of 3

c) Is a multiple of 5.

Sol.: a) 0, 2, 4, 6, 8; b) 1, 4, 7; c) 0, 5.

8. Calculate the prime factorization of:

a) 105   b) 700

Sol.: a) $3 \cdot 5 \cdot 7$; b) $2^2 \cdot 5^2 \cdot 7$.

9. Calculate:

a) LCM(90, 150)   b) LCM(8, 12, 18)

Sol.: a) 450; b) 72.

10. Calculate:

a) HCF(135, 180)   b) HCF (45, 60, 105)

Sol.: a) 540; b) 1260.

11. If a is a multiple of b, what is LCM of a and b?   **Sol.:** a.

12. Find out four pairs multiple-divisor:

| 143 | 12 | 124 | 364 |

| 180 | 31 | 52 | 13 |

13. Write down four consecutive multiples of:

a) 7 greater than 100   b) 15 greater than 230

14. Write down all the multiples of 6 between 92 and 109.

15. Write down all the factors of:

a) 18   b) 50   c) 80

Sol.: a) 1, 2, 3, 6, 9, 18; b) 1, 2, 5, 10, 25, 50;

c) 1, 2, 4, 5, 8, 10, 16, 20, 40, 80.

16. Find out the missing figure so the number (there can be more than one answer):

a) 3[ ]1 is a multiple of 3

b) 57[ ] is a multiple of 2

c) 23[ ] is a multiple of 5

d) 52[ ]3 is a multiple of 11

Sol.: a) 2, 5, 8; b) 0, 2, 4, 6, 8; c) 0, 5; d) 0.

17. We want to distribute 100 liters of water in bottles which all have the same capacity. Find out all the different solutions. Indicate how many bottles we get in each case and the capacity of each one.

18. Sandra can pack her books in boxes of 5, 6 and 9. She has less than 100 books. How many books has she got?   **Sol.:** 9.

19. We want to divide a rectangle of 600 cm by 90 cm into equal squares. Find out the length of the biggest square in cm. Calculate how many squares we get. **Sol.:** 30.

20. In the library of my school there are between 150 and 200 books. Find out, exactly, how many there are, knowing they can be perfectly packed into boxes containing 5, 9, 15 and 18 books each. **Sol.:** 180.

21. Iberia has a flight from Madrid to Ankara every 8 days, British Airways one every 12 days and Easy Jet one every 6 days; one day all three have a flight to Ankara. After how many days will the three flights coincide again? **Sol.:** In 24 days.

22. A group of students can be organized in lines of 5, 4 and 3 students and there are less than 100 students. How many are there? **Sol.:** 60.

23. On a Christmas tree, there are two strings of lights, red lights flash every 24 seconds and green lights every 36 seconds. They start flashing simultaneously when we connect the tree. When will they flash together again? **Sol.:** In 72 seconds.

24. Two friends want to exchange notebooks (3.6 € each) and calculators (4.8 € each). How many products do they have to exchange to have a fair exchange? **Sol.:** 4 and 3.

25. In a group, number of girls is twice the number of boys. How many people are there in the group, 17, 20, 24 or 26? **Sol.:** 24.

JUST AS A CHALLENGE

26. Some friends have gone to the mountain. Every 3 of them are sharing a bag with their meal, every 4, a compass, and every 6, a map. If between bags, compasses and maps, are 27, how many friends did go to the mountain? **Sol.:** 36.

27. Guests to a wedding can be distributed in tables of 3, 5 or 25, but not in tables of 4 or 9. How many guests are there, if we know there are between 1000 and 1250? **Sol.:** 1050.

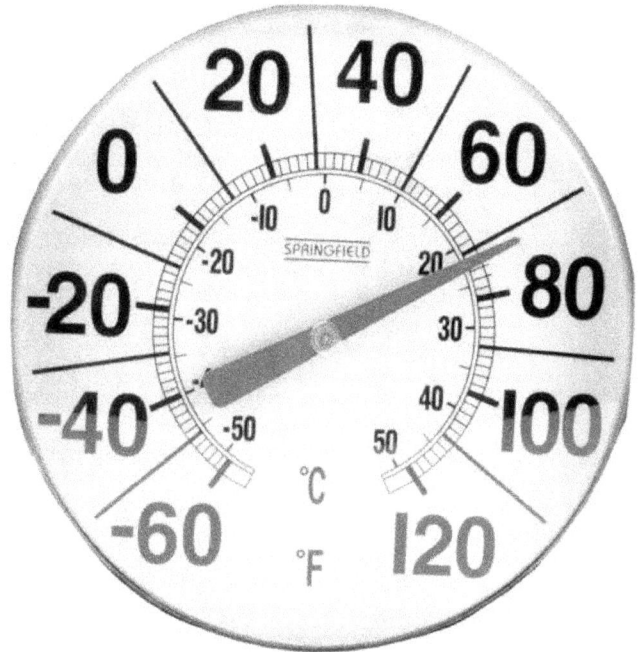

Unit 2.- Integers

1. Negative numbers

Negative numbers are an extremely useful tool for many kinds of problems. For instance, the concept of a negative temperature, the notion of a negative balance in a bank account, and the use of negative numbers to describe the motion of an object in a particular direction are important uses of negative numbers.

The following examples show specific situations where negative numbers are used.

On the Celsius temperature scale, the freezing point of water is 0°C. When we want to represent a temperature in Celsius degrees (written °C) that is colder (less than) the freezing point of water, we represent it by a negative number of degrees. For instance, a temperature that is 7 degrees colder than the freezing point of water on the Celsius scale would be represented using the negative number (-7), and written as (-7) °C.

Note: Negative numbers are usually written into parenthesis. This is very useful for us to be alert when they appear.

When recording the assets of corporations, a deficit (a situation where the corporation owes more money than it has assets) can be described by saying that the company has negative assets. So, for instance, if the corporation has $3,000,000 in assets but owes $4,000,000, then we say the corporation has assets of "negative one million dollars," written as -$1,000,000.

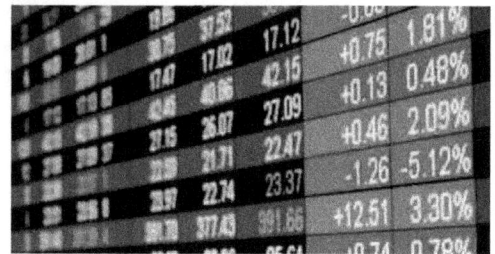

When measuring changes in a quantity, such as a change in temperature or a change in rainfall from one year to the next, we use a negative number to describe a change in which the quantity decreases. For instance, if the temperature at noon is 20°C and the temperature at 6 P.M. is 12°C, we say that the change in temperature from noon to 6 P.M. is -8°C.

Historically, the recognition and use of negative numbers developed very late. Despite the fact that mathematicians in ancient civilizations did perform subtraction, the recognition of negative numbers as "legitimate" numbers did not occur until the 1600s.

The number Line

The number line is a line labeled with the integers in increasing order from left to right, that extends in both directions:

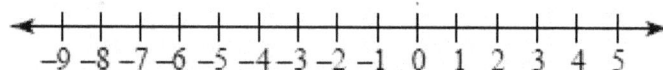

> For any two different places on the number line, the integer on the right is greater than the integer on the left.

For example, $9 > 4$ Is read: nine is 'greater than' four

(-7) < 9   Is read: minus seven is 'less   than' nine.

Exercises

1. Represent on the Number Line the following whole numbers:

$3, -5, 14, -12, -6, 0, 7, -1, 2, -13.$

+++
                                                0

2. List the following temperatures from coolest to hottest:

$3°, -2°, 5°, 2°, -8°, -5°, 6°, 0°, -1°$

3.- Last year, these were the minimum temperatures in Burgos. List the months from the coolest to the hottest.

Ene	Feb	Mar	Abr	May	Jun	Jul	Ago	Sep	Oct	Nov	Dic
$-4°$	$0°$	$-2°$	$6°$	$5°$	$12°$	$16°$	$9°$	$7°$	$2°$	$-1°$	$-6°$

4. Indicate if the following comparisons are true:

a) $3 > -5$ b) $-5 < -7$ c) $-4 < 2$ d) $6 < -1$ e) $0 > -3$ f) $9 < -7$ g) $-2 < -6$ h) $-8 > -5$

5.- Write $<$ or $>$ to make the following comparison true:

a) $-5 \quad 4$ b) $3 \quad -1$ c) $-6 \quad -4$ d) $-3 \quad -2$ e) $0 \quad 3$

2. Opposite and Absolute value of a number

The *opposite* of a number is the number on the number line that is the same distance from zero but on the opposite side. To find the opposite of a number, we only have to change its sign.

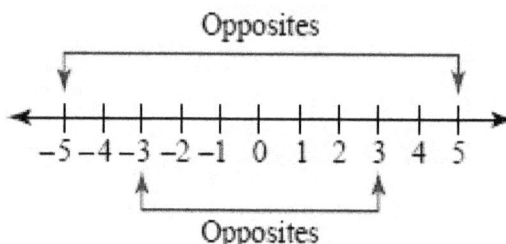

Example 1:

a) The opposite of 26 is (-26)

b) The opposite of (-19) is 19.

The **absolute value** of a signed number is the number of units between the number and zero on the number line. Absolute value is defined as the number of units only; direction is not involved. Therefore, the absolute value is never negative.

- If the number is *positive*, the absolute value is the *same* number.
- If the number is *negative*, the absolute value is the *opposite*.

The absolute value of a number is always a positive number (or zero). We specify the absolute value of a number n by writing n in between two vertical bars: $|n|$.

Example 2: Calculate the absolute value of 15, (– 8), 6, (– 10), 0, 123 and (– 3404):

Solution:

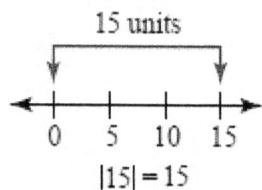

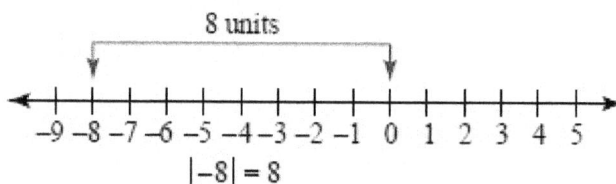

$|6| = 6$ $\qquad$ $|(-10)| = 10$ $\qquad$ $|0| = 0$ $\qquad$ $|123| = 123$ $\qquad$ $|(-3404)| = 3404$

6. Find the opposite of these numbers:

   a) (-42)   b) (-57)   c) 3.78   d) (-0.55)   e) 0.732

7. Find the absolute value of the signed number.

   a) |(-0.25)| b) |-(-6)| c) $|\frac{5}{7}|$ d) $|-\frac{2}{3}|$ e) |0.1657|

8. Simplify:

   a) |16 - 10| - |14 - 9| b) 8 - |12 - 8| - |10 - 8| + 2

---

3. Addition of integers

There is a way to understand how to add integers. In order to add positive and negative integers, we will imagine that we are moving along a number line.

If we want to add (-1) and 5, we start by finding the number -1 on the number line, exactly one unit to the left of zero. Then we would move five units to the right. Since we landed four units to the right of zero, the answer must be 4.

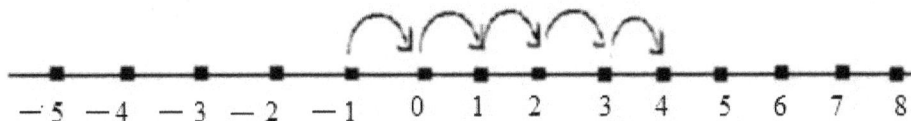

If asked to add 3 and (-5), we can start by finding the number three on the number line (to the right of zero).

Then we move five units left from there because negative numbers make us move to the left side of the number line. Since our last position is two units to the left of zero, the answer is (-2).

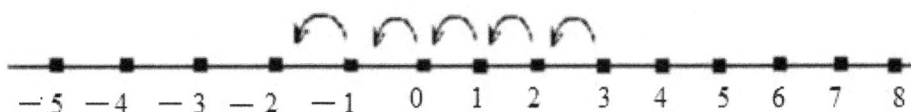

> **To add two integers,**
>
> **1.** If the signs are alike, add their absolute values and use the common sign for the sum.
>
> **2.** If the signs are not alike, subtract the smaller absolute value from the larger absolute value. The sum will have the sign of the number with the larger absolute value.

Example 3.

1. If the signs are alike, add their absolute values and use the common sign for the sum.

    a) $5 + 2 = 7$ b) $(-5) + (-3) = (-8)$ c) $7 + 4 = 11$ d) $(-20) + (-10) = (-30)$

2. If the signs are not alike, subtract the smaller absolute value from the larger absolute value. The sum will have the sign of the number with the larger absolute value.

    a) $3 + (-5) = (-2)$ b) $(-7) + 5 = (-2)$ c) $10 + (-8) = 2$ d) $5 + (-6) = (-1)$

Exercises	**9.** Evaluate: a) $9 - 4$ b) $4 - 9$ c) $10 - 8$ d) $8 - 9$ e) $11 - 7$ f) $7 - 11$ g) $5 - 11$ h) $3 - 7$ i) $1 - 6$ j) $10 - 12$ k) $11 - 15$ l) $14 - 20$ **10.** Calculate: a) $5 - 9$ b) $5 - 11$ c) $13 - 9$ d) $22 - 30$ e) $21 - 33$ f) $46 - 52$ g) $-8 - 14$ h) $-21 - 15$ i) $-33 - 22$ j) $-13 + 18$ k) $-22 + 9$ l) $-37 + 21$

In Maths, it is no allowed to write two signs (+) or (-), together before a number. When this happens, we must change both signs by only one. We will do it as follows:

You must pay attention to the first sign:

- If the first sign is *(+)*, then, quit it and *leave* the second one.
- If the first sign is *(-)*, then, quit it and *change* the second by its opposite.

Example 4:

a) $+(-7) = (-7)$ b) $-(-2) = +2$ c) $-(+8) = (-8)$ d) $-(+1) = (-1)$

e) $+(+11) = +11$ f) $+(-14) = (-14)$

11. Calculate, as shown in the first, made for you:

a) $-(-3) + (-7) = +3 - 7 = (-4)$ b) $3 - (-2)$ c) $15 + (-21)$ d) $(-52) - 24$

e) $2 + (-3)$ f) $-(-3) + (-4)$ g) $(-1) - (-6)$

12. Add:

a) $1 + 3$ b) $(-8) + 5$ c) $2 + 1$ d) $(-4) + 4$ e) $5 + (-6)$

f) $(-6) + 4$ g) $7 + (-2)$ h) $(-3) + 5$ i) $(-3) + (-2)$ j) $(-4) + (-2)$

k) $(-2) + (-2)$ l) $(-5) + 6$ m) $(-2) + (-5)$ n) $(-5) + 10$ ñ) $(-7) + (-4)$

o) $(-4) + 7$ p) $(-5) + (-2)$ q) $(-7) + (-10)$

13. Complete:

a) $(-11) + \boxed{....} = 4$ b) $(+13) + \boxed{....} = 12$ c) $\boxed{....} + (-20) = (-12)$

d) $(+3) + \boxed{....} = (-7)$ e) $(-15) + \boxed{....} = 9$ f) $\boxed{....} + 8 = 7$

5. Addition and subtraction of more than two integers

In real life, in Nature, addition and subtractions are related to more than two numbers. In this part of the chapter, you will learn to calculate them.

When you have to calculate additions and subtractions implying more than two numbers, you must add separately positive numbers, and later, negative numbers. Finally calculate the difference

$$\boxed{\text{TOTAL POSITIVES}} - \boxed{\text{TOTAL NEGATIVES}}$$

To calculate additions and subtractions implying more than two numbers:

1. Add separately positive numbers,

2. Add separately negative numbers,

3. Finally, calculate the difference

$$\boxed{\text{TOTAL POSITIVES}} - \boxed{\text{TOTAL NEGATIVES}}$$

Example 5:   Calculate:

   a) $7 - 2 + 4 + 1 - 8$ 　　　　 b) $10 + 2 - 5 - 9 + 2$ 　　　　 c) $-2 + 5 - 4 - 10 + 1$

Solution:

a) $5 - 2 + 4 + 1 - 8$

   1st: add separately positive numbers $\rightarrow +7 + 4 + 1 = 12$

   2nd: add separately negative numbers $\rightarrow -2 - 8 = -10$

   3rd: finally, calculate the difference $\rightarrow 12 - 10 = \boxed{2}$

b) $10 + 2 - 5 - 9 + 2$

   1st: add separately positive numbers $\rightarrow +10 + 2 + 2 = 14$

   2nd: add separately negative numbers $\rightarrow -5 - 9 = -14$

   3rd: finally, calculate the difference $\rightarrow 14 - 14 = \boxed{0}$

c) $-2 + 5 - 4 - 10 + 1$

1st: add separately positive numbers $\rightarrow +5 + 1 = 6$

2nd: add separately negative numbers $\rightarrow -2 - 4 - 10 = -16$

3rd: finally, calculate the difference $\rightarrow 6 - 16 = \boxed{(-10)}$

14. Calculate the following sums and subtractions of integers:

a) $5 + 2 - 4 - 4 + 2$ b) $3 - 5 + 6 - 1$ c) $-2 + 1 - 2 + 7$

d) $10 + 1 + 2 - 7 + 2$ e) $-1 - 4 - 2$ f) $-3 - 12 + 3 + 9 - 4$

g) $5 - 3 - 7 + 1 + 8$ h) $2 - 3 + 4 + 1 - 8 + 2$ i) $1 - 3 + 5 - 7 + 9 - 11$

j) $2 + 4 - 6 - 8 + 10 - 12 + 14$

15. Opera:

a) $-3 - 5 + 4 - 1 + 3$ b) $34 - 23 + 41 + 23 - 35$ c) $12 - 4 + 6 + 7 - 3 - 3$

d) $5 - 2 + 6 - 11 - 3 - 5 + 8 - 1$ e) $-3 - 4 - 3 + 10 + 1 - 1$ f) $12 + 3 - 15 + 9 - 8$

16. Calculate:

a) $23 - 4 - 5 - 1 + 2 - 20$ b) $-2 - 2 - 2 - 2 - 2$

c) $4 - 6 + 4 - 6 - 6 + 4 + 4 - 6 - 6$ d) $12 - 3 - 1 - 2 - 4$

e) $-2 + 1 + 1 - 1 + 4 + 2 - 3$ f) $23 + 4 - 456 - 23 + 2 + 456 - 7$

17. Calculate:

a) $4 + (-3)$ b) $8 - (-5)$ c) $4 + (-2) - (-3) + 5$

d) $5 - (-2) + 6 - (-4)$ e) $(-5) - (-3) + 10 - (-2)$ f) $4 + (-7) - (-2) - 3$

g) $(-8) + (-3) - (-1) + (-4)$ h) $(-9) - 12 - 1$

6. Addition and subtraction with parentheses

In operations implying parenthesis, pay attention:

FIRST OPERATE INTO THE PARENTHESIS!!!

Example 6: Calculate:

    a) $13 - (6 + 5)$ b) $(4 + 8) - (3 - 9)$ c) $10 + (8 - 15 + 2 - 6)$

Solutions:

a) $13 - (6 + 5)$

    First, operate into the parentheses: $(6 + 5) = 11$

    Then, substitute in the main operation line: $13 - 11 = \boxed{2}$.

b) $(4 + 8) - (3 - 9)$

    First, operate into the parentheses: $\begin{cases} (4+8) = 12 \\ (3-9) = (-6) \end{cases}$

    Then, substitute in the main operation line: $12 - (-6) = 12 + 6 = \boxed{8}$.

c) $10 + (8 - 15 + 2 - 6)$

    First, operate into the parentheses: $(8 - 15 + 2 - 6) =$

                                1st: add separately positive numbers $\rightarrow +8 + 2 = 10$

                                2nd: add separately negative numbers $\rightarrow -15 - 6 = -21$

                                3rd: finally, calculate the difference $\rightarrow 10 - 21 = (-11)$

    Then, substitute in the main operation line: $10 + (-11) = 10 - 11 = \boxed{(-1)}$.

18. Remove parentheses and calculate:

a) $1 - (7 - 2 - 10) - (3 - 8)$

b) $(8 - 4 - 3) - (5 - 8 - 1)$

c) $(3 - 5) - (1 - 4) + (5 - 8)$

d) $3 - (5 - 8) - (11 - 4) + (13 - 9)$

e) $5 - (7 - 2 - 10) - (3 - 8)$

f) $(8 - 2 - 3) - (5 - 8 - 1)$

g) $(2 - 5) - (1 - 4) + (5 - 8)$

h) $10 - (5 - 8) - (11 - 4) + (13 - 9)$

19. Calculate:

a) $-18 - (-9)$

b) $15 - (-10 + 18)$

c) $-12 - (-12 - 4)$

d) $- (-12) - (12 + 7)$

e) $-7 + 32$

f) $-12 + 8 + (-10) + 4$

g) $- (-8) + 12 + (-5) + 7$

h) $-15 + 30 + (-19) + 20$

20. Work out:

a) $-15 + 7 + (-8) + 9$

b) $15 + (-10) + 5 + (-10)$

c) $-12 + 8 + (-7) + (-2)$

d) $-13 + 15 + (-10) + 9$

e) $-20 - (-8 + 4 - 5)$

f) $12 - (-8 + 10)$

g) $- (-8) - (4 - 7 - 9)$

h) $-15 - (-2 + 9)$

7. Multiplying and dividing integers

When multiplying or dividing integers, we will multiply or divide their absolute values (in the same way as when multiplying or dividing natural numbers). But, after that, we will have to pay attention to their signs, because:

Product of integers of the **same sign** will give a **positive number**.

Product of integers with **opposite sign** will give a **negative number.**

If any of the integers in the product is 0, the product is 0.

* And, the same rules about signs are applied in divisions.

Example 7: Calculate:

a) 4×3 b) $(-4) \times (-5)$ c) $(-7) \times 6$ d) $12 \times (-2)$

e) $14 \div 2$ f) $(-24) \div (-3)$ g) $(-100) \div 25$ h) $98 \div (-7)$

Solution:

a) 4×3 Both numbers are positive (same sign), so their product is a positive number, 12.

b) $(-4) \times (-5)$ Both numbers are negative (same sign), so their product is a positive number, 20.

c) $(-7) \times 6$, the first number is negative and the second is positive, so the product is a negative number, (-42).

d) $12 \times (-2)$ the first number is positive and the second is negative, so the product is a negative number, (-24).

e) $14 \div 2$ Both numbers are positive, so the quotient is a positive number, 7.

f) $(-24) \div (-3)$ Both numbers are negative, so the quotient is a positive number, 8.

g) $(-100) \div 25$ The numbers have different signs, so the quotient is a negative number, (- 4).

h) $98 \div (-7)$ Both numbers have different signs, so the quotient is a negative number, (-14).

Exercises

21. Calculate the following products:

a) $5 \cdot 4$ b) $(-3) \cdot (-8)$ c) $3 \cdot (-5)$ d) $(-4) \cdot 5$ e) $(-7) \cdot (-6)$

f) $5 \cdot 7$ g) $(-4) \cdot 8$ h) $9 \cdot (-5)$ i) $(-11) \cdot 7$ j) $10 \cdot (-10)$

k) $(-8) \cdot 5$ l) $(-10) \cdot (-10)$

22. Calculate the following products:

a) $(-3) \cdot (-5)$ b) $(-4) \cdot (-2)$ c) $9 \cdot (-6)$ d) $6 \cdot (-3)$ e) $9 \cdot (-9)$

f) $5 \cdot (-8)$ g) $(-5) \cdot (-8)$ h) $(-7) \cdot 9$ i) $9 \cdot (-5)$ j) $(-4) \cdot (-7)$

k) $(-5) \cdot 5$ l) $(-10) \cdot (-2)$

23. Calculate the following products:

a) $(+21) \cdot (+3) \cdot (+4)$ b) $(+19) \cdot (-2) \cdot (+3)$ c) $(+13) \cdot (-5) \cdot (-6)$

d) $(-20) \cdot (-9) \cdot (-3)$

24. Copy and complete:

a) (-5) · [.....] = (-30)   b) [.....] · (+3) = 45   c) (-9) · [.....] = 27

d) [.....] · (-8) = -48

25. Divide:

a) (− 15) : (- 5)   b) (− 4) : (−2)   c) 9 : (− 3)   d) 6 : (− 3)

e) 9 : (− 9)   f) 16 : (− 8)   g) (-40) : (− 8)   h) (- 63) : 9

26. Divide:

a) 25 : (− 5)   b) (− 21) : (− 7)   c) 8 : (− 4)   d) 4 : (− 1)

e) 8 : (− 4)   f) (− 24) : (− 3)   g) (+ 12) : 2   h) 72 : (− 8)

27. Copy and complete:

a) [.....] : 4 = (-10)   b) 40 : [.....] = (-8)   c) (-100) : [.....] = (-25)

d) [.....] : (-12) = 6

28. Copy and complete:

a) (-36) : [.....] = (- 4)   b) (-54) : [.....] = 9   c) [.....] : (-6) = (-42)

d) 48 : [.....] = (- 6)   e) (-63) : [.....] = (-7)   f) [.....] : 8 = 2

29. Calculate these products and quotients:

a) 4 · (−3)   b) (−5) · 2   c) 3 · (−7)   d) (−2) · (−5)

e) 3 · (−4)   f) (−5) · 3   g) 4 · (−2) · 3   h) (−3) · (−2) · 4

i) (−1) · (−4) · (−5)   j) (−7) · 2 · (−3)   k) (−2) · 4 · (−3)   l) (−12) : 4

m) 24 : (−6)   n) (−8) : (−2)   ñ) (−27) : (−1)   o) (−12) : (−4)

p) 18 : (−3)   q) (−6) · (−2) : (−4)   r) 3 · (−8) : (+6)   s) 6 · (−8) : (−12)

t) (−7) · 6 : (−21)   u) 4 · (−9) : (+12)

30. At 7 am the temperature was 4 degrees below zero, five hours later showed 3 ° C above zero. What is the difference between the two temperatures?

31. Mary lives in 3rd floor. She goes down 5 floors to go to the storeroom and then rises 7 to visit his friend Alberto. What floor is Alberto living at?

32. Sara leaves her car in her parking, located in the third basement and goes up 4 floors to her house. What floor is she living at?

33. On the 1st of December, the level of the water in a reservoir was 130 cm above its average level. On the 1st of July it was 110 cm below its average level. How many cm did the water level drop in this time?

8. Order of operations

To evaluate an expression with more than one operation

Step 1. Parentheses

Step 2. Exponents

Step 3. Multiply and Divide

Step 4. Add and Subtract

Example 8: Calculate:

a) $2 \cdot 7 - 3 \cdot 4 - 2 \cdot 3$ b) $30 : 6 - 42 : 7 - 27 : 9$ c) $16 + (-5) \cdot 4$ d) $20 - (-6) \cdot (-4)$

Solution:

a) $2 \cdot 7 - 3 \cdot 4 - 2 \cdot 3$ →   No parentheses, no exponents, so, we go directly to step 3, multiplications:

$$2 \cdot 7 = 14$$
$$3 \cdot 4 = 12$$
$$2 \cdot 3 = 6$$

And substitute: $14 - 12 - 6 = 14 - 18 = \boxed{(-4)}$.

b) $30 : 6 - 42 : 7 - 27 : 9$ $\rightarrow$ No parentheses, no exponents, so, we go directly to step 3, divisions: $\qquad\qquad\qquad 30 : 6 = 5$

$$42 : 7 = 6$$

$$27 : 9 = 3$$

And substitute: $5 - 6 - 3 = 5 - 9 = \boxed{(- 4)}$.

c) $16 + (–5) \cdot 4$ $\rightarrow$ No parentheses, no exponents, so, we go directly to step 3, multiplication:

$$(-5) \cdot 4 = (-20)$$

And substitute: $16 + (-20) = \boxed{(- 4)}$.

d) $20 - (– 6) \cdot (–4)$ $\rightarrow$ No parentheses, no exponents, so, we go directly to step 3, multiplication:

$$(- 6) \cdot (- 4) = 24$$

And substitute: $20 - 24 = \boxed{(- 4)}$.

---

34. Calculate. Pay attention to order of operations:

a) $3 \cdot 5 - 4 \cdot 6 + 5 \cdot 4 - 6 \cdot 5$ $\qquad$ b) $5 \cdot 4 - 28 : 4 - 3 \cdot 3$

c) $(–2) \cdot (–5) + (+4) \cdot (–3)$ $\qquad$ d) $(–8) \cdot (+2) - (+5) \cdot (–4)$

e) $10 + (– 4) \cdot (+2) - (+6)$ $\qquad$ f) $(–5) - (+4) \cdot (–3) - (–8)$

g) $14 - (+5) \cdot (– 4) + (– 6) \cdot (+3) + (–8)$ $\qquad$ h) $(+4) \cdot (– 6) - (–15) - (+2) \cdot (–7) - (+12)$

Exercises

Example 9: Calculate:

a) $60 : (8 – 14) + 12$ $\qquad$ b) $(9 – 13 – 6 + 9) \cdot (5 – 11 + 7 – 4)$

Solutions:

a) $60 : (8 – 14) + 12$

Step 1. Parentheses $\rightarrow$ $8 – 14 = (- 6)$

Step 2. Exponents $\rightarrow$ There are not. $\rightarrow$ So we have: $60 : (- 6) + 12$

Step 3. Division $\rightarrow$ $60 : (- 6) = (-10)$

Step 4. Add and Subtract $\rightarrow$ $(-10) + 12 = \boxed{(- 2)}$.

b) $(9 - 13 - 6 + 9) \cdot (5 - 11 + 7 - 4)$

Step 1. Parentheses $\rightarrow$ $\begin{cases} (9 - 13 - 6 + 9) = 18 - 19 = (-1) \\ (5 - 11 + 7 - 4) = 12 - 15 = (-3) \end{cases}$

Step 2. Exponents $\rightarrow$ There are not. $\rightarrow$ So we have: $(-1) \cdot (-3)$

Step 3. Product $\rightarrow$ $(-1) \cdot (-3) = \boxed{3}$.

---

35. Calculate. Pay attention to order of operations:

a) $15 + 2 \cdot 3 - 8 : 2$

b) $4 + 2 \cdot 3$

c) $(4 + 2) \cdot 3$

d) $4 + (2 \cdot 3)$

e) $10 - 1 \cdot 3 + 4$

f) $3 + 2 \cdot 5 - 6 : 3$

g) $3 - 10 : 5 + 6 \cdot 3 - 2$

h) $3 \cdot (5 - 3) + 5 \cdot 4 + 10$

i) $5 - 3 \cdot (4 - 3) + 4 \cdot (13 - 3)$

36. Calculate:

a) $12 + 6 - 4 + 4 - 7$

b) $4 \cdot 5 + 3 - 16 : 2 + 4$

c) $2 \cdot 5 + 3 \cdot 12 : 4 - 20$

d) $15 - 2 \cdot 5 + (4 + 16) : 5$

e) $13 - (2 + 1) : 3 + (2 + 3) \cdot 2$

f) $10 \cdot (3 + 2) - 5 \cdot 4 + 5 - 10 : 2$

g) $3 \cdot (4 + 1) \cdot 2 + 1 \cdot (14 - 5 \cdot 2)$

h) $10 + 2 \cdot (7 - 5) - 10$

37. Calculate:

a) $16 + (-5) \cdot (+4)$

b) $20 - (-6) \cdot (-4)$

c) $(-2) \cdot (-5) + (+4) \cdot (-3)$

d) $(-8) \cdot (+2) - (+5) \cdot (-4)$

38. Calculate:

a) $(+45) : [(-7) + (+2)]$

b) $(+2) \cdot [(-63) : (-7)]$

c) $(-25) : [(+3) - (+8)]$

d) $(-8) \cdot [(+21) : (-3)]$

e) $(-7) - [(-14) : (+2) - (-7)]$

Exercises

A **Power** is a number obtained by multiplying a number by itself a certain number of times

$$a^n = \underbrace{a \cdot a \cdot a \cdot a \ldots a}_{n \ times} \quad where \quad \begin{cases} a : base \\ b : exponent \end{cases}$$

For example, $6^5 = 6 \cdot 6 \cdot 6 \cdot 6 \cdot 6$, and can be read as:

- Six to the fifth power

- Six to the power of five

- Six powered to five.

The most common is **six to the power of five.**

The number that is successively multiplied by itself is called the **base**. A small raised number called the **exponent** follows the base and indicates the number of times the base is to be multiplied.

And, remember: $a^1 = a$ and $a^0 = 1$, for any natural number a.

Especial cases: **Squares and cubes**

Although 3^2 is read as three to the second power, three to the power of two, three to the square power, the most common is **three squared.**

Although 5^3 is read as five to the third power, five to the power of three, the most common is **five cubed.**

Example 10: What is the largest square you can form with 36 cards? And with 52? Will you have any left?

Solution: $1^2 = 1$; $2^2 = 4$; $3^2 = 9$...... $6^2 = 36$; $7^2 = 49$; $8^2 = 64$.   So, With 36 cards, you can exactly form a 6 cards sided square, and with 52, a 7 cards sided, in this case, leaving 3 cards left ($52 - 49 = 3$).

9.1. Product of powers with the same base

When powers with the same base are multiplied, the base remains unchanged and the **exponents are added.**

$$a^m \cdot a^n = a^{(m+n)}$$

39. Write as an only power:

a) $(-4)^4 \cdot (-4)^{10}$ b) $5^2 \cdot 5^{(-6)}$ c) $7^{(-4)} \cdot 7^{(-5)}$ d) $5^{(-4)} \cdot 5^6$ e) $7 \cdot 7^{(-2)}$

f) $4^{(-5)} \cdot 4^{(-7)}$ g) $(-5)^6 \cdot (-5)^{(-8)} \cdot (-5)^2 \cdot (-5)$

40. Find the value of x that makes the following expressions true:

a) $(-8)^x \cdot (-8)^4 = (-8)^{10}$ b) $3^2 \cdot 3^x \cdot 3^4 = 3^3$ c) $(-2)^8 \cdot (-2)^x \cdot (-2)^5 = (-2)^4$

d) $(-3)^4 \cdot (-3)^{(-10)} \cdot (-3)^8 \cdot (-3)^2 = (-3)^x$ e) $(-2)^4 \cdot (-2)^{(-5)} \cdot (-2) = (-2)^x$

9.2. Quotient of powers with the same base

When powers with the same base are divided, the base remains unchanged and the **exponents are subtracted.**

$$\frac{a^m}{a^n} = a^{(m-n)}$$

41. Write as an only power:

a) $(-5)^5 : (-5)^2$ b) $(-3)^3 : (-3)^7$ c) $2^{-5} : 2^4$ d) $5^2 : 5^{-2}$

e) $5^{-3} : 5^2$ f) $(-7)^{-5} : (-7)^{-4}$ g) $8^{-5} : 8^{-2}$ h) $(-2)^{-5} : (-2)^3$

42. Find the value of x that makes the following expressions true:

a) $(-8)^x : (-8)^4 = (-8)^{10}$ b) $3^2 : 3^x = 3^3$ c) $(-2)^x : (-2)^5 = (-2)^4$

d) $(-3)^4 \cdot (-3)^{(-10)} : (-3)^8 = (-3)^x$ e) $(-2)^x \cdot (-2)^{(-5)} \cdot (-2) = (-2)$

9.3. Power of a power

To calculate a power of a power, we leave the same base, while the exponents must be **multiplied.**

$$\left(a^m\right)^n = a^{m \cdot n}$$

43. Calculate:

a) $((-2)^3)^2$ b) $(3^{-2})^4$ c) $((5^{-3})^{-2})^{-4}$ d) $((-5)^{-2})^{-4}$ e) $(7^{-2})^0$ f) $((8^{-3})^0)^{-4}$

44. Fill in the missing numbers:

a) $(4^2)^{-5} = 4^{[\ ]}$ b) $(3^2)^{[\ ]} = 3^{-10}$ c) $((-7)^{[\ ]})^{-3} = (-7)^{15}$

d) $([\ ]^2)^3 = (-10)^6$ e) $((-2)^2)^{[\ ]} = ([\ ])^2$.

45. By using previous rules, express the following as an only power:

a) $7^3 \cdot 7^2$ b) $8^{10} \cdot 8^{20} \cdot 8^1$ c) $4^5 : 4^2$ d) $\left(3^2\right)^6$

e) $10^7 : 10^4$ f) $\left(9^2\right)^8$ g) $10^4 \cdot 10^2 \cdot 10^5$ h) $\left(3^3 \cdot 3^6\right) : 3^7$

46. Express the following as an only power:

a) $(-3)^2 \cdot (-3)^{-3}$ b) $6^{-2} : 6^{-6}$ c) $x^2 : x^{-6}$ d) $x^{-2} : x^6$ e) $\left(2^{-3}\right)^4$

f) $\left(5^2\right)^{-4}$ g) $\left(x^{-3}\right)^{-2}$ h) $\left(x^2 y^3\right)^2$ i) $5^{-2} : 5^{-3}$

47. Express the following as an only power:

a) $6^{-2} \cdot 6^0 =$ b) $2^{-4} \cdot 2^6 =$ c) $3^{-2} \cdot 3^{-2} =$ d) $5^0 : 5^{-2} =$

e) $6^{-3} : 6^0 =$ f) $5^8 : 5^{-4} =$ g) $6^{-3} : 6^{-7} =$ h) $\left(4^0\right)^5 =$

9.4. Powers with different bases but the same exponent

> **Product:** When powers **with the same exponent** are multiplied, multiply the bases and keep the same exponent.
>
> $$(a \cdot b)^m = a^m \cdot b^m$$

> **Quotient:** When powers **with the same exponent** are divided, bases are divided and the **exponent remains unchanged.**
>
> $$\left(\frac{a}{b}\right)^m = \frac{a^m}{b^m}$$

Exercises

48. Complete:

a) $3^7 \cdot 8^7 = [\ ]^7$

b) $[\ ]^2 \cdot [\ ]^2 = 6^2$

c) $5^2 \cdot [\ ]^2 = 15^2$

d) $2^2 \cdot [\ ]^2 = 14^2$

e) $8^5 : 4^5 = [\ ]^5$

f) $\dfrac{15^5}{5^5} = [\ ]^5$

g) $\dfrac{16^7}{8^7} = [\ ]^7$

h) $\dfrac{6^{12}}{3^{12}} = [\ ]^{[\ ]}$

49. Write as an only power:

a) $5^4 \cdot 3^4$
b) $2^2 \cdot 4^2$
c) $3^4 \cdot 4^4 \cdot 5^4$
d) $x^n \cdot y^n$
e) $(-5)^4 \cdot (-3)^4$

f) $8^3 : 4^3$
g) $9^5 : 3^5$
h) $x^n : y^n$
i) $(-10)^4 : 2^4$
j) $(-6)^3 : (-3)^3$

h) $\left(4^0\right)^5 =$

In the next table, you are shown some examples that should help you to understand and remember the rules of powers.

Laws	Examples
$a^m \cdot a^n = a^{(m+n)}$	$a^4 \cdot a^3 = (a \cdot a \cdot a \cdot a) \cdot (a \cdot a \cdot a) = a \cdot a \cdot a \cdot a \cdot a \cdot a \cdot a = a^7$
$\dfrac{a^m}{a^n} = a^{(m-n)}$	$\dfrac{a^8}{a^3} = \dfrac{a \cdot a \cdot a \cdot a \cdot a \cdot \cancel{a} \cdot \cancel{a} \cdot \cancel{a}}{\cancel{a} \cdot \cancel{a} \cdot \cancel{a}} = a \cdot a \cdot a \cdot a \cdot a = a^5$
$\left(a^m\right)^n = a^{m \cdot n}$	$(a^4)^2 = a^4 \cdot a^4 = a \cdot a \cdot a \cdot a \cdot a \cdot a \cdot a \cdot a = a^8$
$(a \cdot b)^m = a^m \cdot b^m$	$(a \cdot b)^3 = (a \cdot b) \cdot (a \cdot b) \cdot (a \cdot b) = a \cdot b \cdot a \cdot b \cdot a \cdot b = (a \cdot a \cdot a) \cdot (b \cdot b \cdot b) = a^3 \cdot b^3$
$\left(\dfrac{a}{b}\right)^m = \dfrac{a^m}{b^m}$	$\left(\dfrac{a}{b}\right)^3 = \dfrac{a}{b} \cdot \dfrac{a}{b} \cdot \dfrac{a}{b} = \dfrac{a \cdot a \cdot a}{b \cdot b \cdot b} = \dfrac{a^3}{b^3}$

Powers having an integer exponent

Till now, we have only studied powers having exponents > 1. This might make you wonder if powers like 5^1, 8^0 or 7^{-2} exist. Answer is affirmative. They exist and their values are:

$$a^1 = a \qquad a^0 = 1 \qquad a^{-n} = \frac{1}{a^n} \qquad \left(\frac{a}{b}\right)^{-n} = \left(\frac{b}{a}\right)^n$$

In the next lines, we are going to demonstrate these values:

Exponent = 1

Choose two numbers whose difference is 1. For example, 4 and 3.

$$a^1 = a^{4-3} = \frac{a^4}{a^3} = \frac{a \cdot \cancel{a} \cdot \cancel{a} \cdot \cancel{a}}{\cancel{a} \cdot \cancel{a} \cdot \cancel{a}} = a \qquad \boxed{a^1 = a}$$

Exponent = 0

Choose two numbers whose difference is 0. For example, 3 and 3.

$$a^0 = a^{3-3} = \frac{a^3}{a^3} = \frac{\cancel{a} \cdot \cancel{a} \cdot \cancel{a}}{\cancel{a} \cdot \cancel{a} \cdot \cancel{a}} = 1 \qquad \boxed{a^0 = 1}$$

Negative exponent

Choose two numbers whose difference is negative. For example 3 and 5 $(3 - 5 = -2)$. $\Rightarrow$ $a^{-2} = a^{3-5} = \dfrac{a^3}{a^5} = \dfrac{\not{a}\cdot\not{a}\cdot\not{a}}{a\cdot a\cdot\not{a}\cdot\not{a}\cdot\not{a}} = \dfrac{1}{a^2}$ $\Rightarrow$ $\boxed{a^{-2} = \dfrac{1}{a^2}}$

And also: $a^{-1} = \left(\dfrac{a}{1}\right)^{-1} = \left(\dfrac{1}{a}\right)^{1} = \dfrac{1}{a}$

Example 11: Calculate: a) 7^{-2} b) $5^{-3}\cdot 5^3$ c) $2^{-2}\cdot 3^{-2}$ d) $(2^3)^{-2}$ e) $(-6)^{-3}$
                             f) $8^2 : 8^{-2}$ g) $12^{-3} : 4^{-3}$

Solution:

a) $7^{-2} = \dfrac{1}{7^2} = \dfrac{1}{49}$ b) $5^{-3}\cdot 5^3 = 5^{-3+3} = 5^0 = 1$

c) $2^{-2}\cdot 3^{-2} = (2\cdot 3)^{-2} = 6^{-2} = \dfrac{1}{6^2} = \dfrac{1}{36}$ d) $(2^3)^{-2} = 2^{-6} = \dfrac{1}{2^6} = \dfrac{1}{64}$

e) $(-6)^{-3} = \dfrac{1}{(-6)^3} = \dfrac{1}{-216} = -\dfrac{1}{216}$ f) $8^2\cdot 8^{-2} = 8^{2-2} = 8^0 = 1$

g) $12^{-3} : 4^{-3} = (12/4)^{-3} = 3^{-3} = \dfrac{1}{3^3} = \dfrac{1}{27}$

Exercises

50. Write as an only power:

a) $7^3\cdot 7^{-2}$ b) $3^5\cdot 3^{-4}\cdot 3^4$ c) $(-5)^2\cdot(-5)^{-5}$ d) $4^{-6} : 4^{-2}$ e) $7^{-1} : 7^3$

f) $x^{-9}\cdot x^5$ g) $(5^2)^{-3}$ h) $((-2)^{-2})^{-3}$ i) $(-2)^9 : (-2)^{-5}$ j) $(-9)^{-7} : (-9)^{-4}$

51. Write as an only power:

a) $(-3)^4\cdot(-5)^4$ b) $3^4\cdot(-2)^4\cdot 5^4$ c) $x^n\cdot y^n$ d) $(-8)^7\cdot 5^7$ e) $(-8)^3 : 4^3$

f) $(-9)^5 : (-3)^5$ g) $x^n : y^n$ h) $(-10)^{-7} : 2^{-7}$ i) $(-6)^5 : 2^5$

52. Substitute the symbol (Δ) by a number:

a) $2^{\Delta}\cdot 2^7 = 2^{-2}$ b) $(-5)^4 : (-5)^{\Delta} = (-5)^{-3}$ c) $(-7)^3 \times (-7)^{-5} = (-7)^{\Delta}$

d) $[a^7]^{\Delta} = a^{-21}$ e) $(8^{-11})^{-5} = 8^{\Delta}$

Till now, we have calculated and worked with powers. In this part of the unit, we a are going to work with the function "root", which is reciprocal to a power.

For example, if we say that $\sqrt{9} = 3$, that means that $3^2 = 9$.

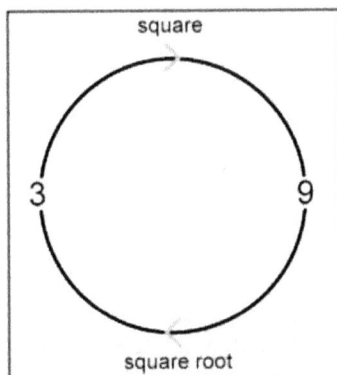

Examples:

$\sqrt{1} = 1$ because $1^2 = 1$
$\sqrt{4} = 2$ because $2^2 = 4$
$\sqrt{9} = 3$ because $3^2 = 9$
$\sqrt{16} = 4$ because $4^2 = 16$
$\sqrt{25} = 5$ because $5^2 = 25$
$\sqrt{36} = 6$ because $6^2 = 36$
$\sqrt{49} = 7$ because $7^2 = 49$

Or, in another way:

POWERS	ROOTS
A square has a side of 5 cm. What is its area? $A = s^2 = 5^2 = 25$ cm^2.	The area of a square is 49 cm^2. What is the length of its side? s = 7 cm, because $7^2 = 49$. This is a root, a square root: $\sqrt{49} = 7 \iff 7^2 = 49$
A cube has a side of 4 m. What is its volume? $V = s^3 = 4^3 = 64$ m^3.	The volume of a cube is cube 8 cm^3. What is the length of its side? s = 2 cm, because $2^3 = 8$. This is a root, a cubic root: $\sqrt[3]{8} = 2 \iff 2^3 = 8$

Root of a real number **a**, written $\sqrt[n]{a}$, where **n** is a natural number, is another real number, **b**, so that

$$\sqrt[n]{a} = b \iff b^n = a$$

In this expression, **n** is named **index** and **a** is named **radicand**.

Number of roots

Even index:

- If radicand is a **positive** number, there are **two opposite roots**.

      For example, squared root of 25 is 5 and (-5).

- If radicand is **0**, its root is **0.**

- If radicand is a **negative** number, it has **no roots,** because any number to an even power, is always a positive number.

      For example, (-4) has no roots, because there is not any number b, so that b2 = (-4).

Odd index:

In this case, all numbers have **an only root**: **Positive** if radicand is **positive**, and **negative** if radicand is **negative**, and **null** if radicand is **null**.

For example, $\sqrt[3]{8} = 2$, $\sqrt[3]{-8} = -2$ and $\sqrt[3]{0} = 0$

Following chart resumes above information:

Index: n	Radicand: a	Number of roots
Even	$a > 0$	2 opposite roots
	$a = 0$	1 null root
	$a < 0$	No roots
	$a > 0$	1 positive root
	$a = 0$	1 null root
	$a < 0$	1 negative root

Exact and whole roots

Sometimes, when calculating squared root of a number, we exactly obtain the solution, like $\sqrt{25} = 5$.

This is named an **exact root**.

But, it is more usual not to obtain an exact solution, but having a remainder. For example, when calculating squared root of 40, we try squares: $3^2 = 9$, $4^2 = 16$, $5^2 = 25$, $6^2 = 36$, $7^2 = 49$.

As $6^2 < 40 < 7^2$, we say 40 has a whole squared root of 6, having a remainder of $40 - 36 = 4$.
So, we say 40 has a **whole square root** of 6 and a **remainder** of 4.

Example 12: Calculate whole root and remainder of 11, 29 and 37.

Solution:

a) $2^2 = 4$, $3^2 = 9$, $4^2 = 16 > 11$ So, whole root is 3 and remainder is $11 - 3^2 = 11 - 9 = 2$.

b) Whole root = 5,   Remainder = $29 - 5^2 = 29 - 25 = 4$.

c) Whole root = 6, Remainder = $37 - 6^2 = 37 - 36 = 1$.

Exercises

53. Calculate the following square roots:

$\sqrt{9}$	$\sqrt{36}$	$\sqrt{-1}$
$\sqrt{-4}$	$\sqrt{100}$	$\sqrt{16}$
$\sqrt{4}$	$\sqrt{49}$	$\sqrt{81}$
$\sqrt{0}$	$\sqrt{64}$	$\sqrt{-9}$
		$\sqrt{1}$

54. Calculate the square root of these numbers:

a) 64   b) 121   c) 144   d) 196

---

55. Calculate, without using pencil and paper, the square root and the remainder of these numbers:

   a) 93   b) 59   c) 130   d) 111

56. Calculate the remainder:

   a) Root = 12   b) Root = 23   c) Root = 30   d) Root = 32

   Radicand = 160   Radicand = 5   Radicand = 901   Radicand = 1030

57. Calculate, without using pencil and paper, which of these statements are false:

   a) $\sqrt{23} = 4$; remainder 7   b) $\sqrt{30} = 5$; remainder 10   c) $\sqrt{45} = 7$; remainder 4

   d) $\sqrt{60} = 7$; remainder 11   e) $\sqrt{80} = 9$; remainder 1   f) $\sqrt{85} = 9$; remainder 5

   g) $\sqrt{96} = 9$; remainder 15   h) $\sqrt{204} = 14$; remainder 2

58. Write all the two digits whole numbers whose whole square root has a remainder of 2.

59. Write all the three digits whole numbers, being less than 500 and whose square roots have a remainder of 10.

60. A number has a whole square root of 5 and its remainder is the biggest it can have. What is the remainder? And the number?

Review exercises

1. Write an integer to represent each description:
   a) Deposit $1,821 into a bank account.   b) A loss of $38,323 on an investment.
   c) 14 units to the right on a number line.   d) 8 units to the left on a number line.
   e) Seventy-six feet above sea level.   f) The stock market went up 145 points today.
   g) One hundred two feet below sea level.   h) The football player had a 20 yard loss on the play.
   i) A loss of six pounds.

2. List the following integers from least to highest: (-3), 5, (-4), 0, 4, (-2).

Sol.: $(-4) < (-3) < (-2) < 0 < 4 < 5$.

3. Ordena de mayor a menor: 4, (-3), 8, (-7), 1, (-9).

Sol.: $8 > 4 > 1 > (-3) > (-7) > (-9)$.

4. Write in the Number Line the integers being bigger than (-4) and less than 6. What is¡s the least of them? How can you find it in the Number Line?

5. Calculate: a) |-3| + |+3| b) |-8| + |-1| c) |-7| - |-4| d) |+8| - |-12| e) |-2| - |-2|

Sol.: a) 6; b) 9;   c) 3;   d) (-4);   e) 0.

6. Insert the symbol "bigger than", "less than" or "equals" in the square:

a) +5 ☐ +11   b) -3 ☐ -7   c) 0 ☐ -6   d) |-8| ☐ |+2|

e) -4 ☐ +3   f) +9 ☐ 0   g) |-5|☐ +5   h) |-9| ☐ |+9|

Sol.: a) <; b) >; c) >; d) >; e) <; f) >; g) =; h) =.

7. Work out:

a) $-2 + 6$ b) $-4 + 7$ c) $-1 + 9$ d) $-7 + 2$ e) $-8 + 5$

f) $-10 + 8$ g) $-12 + 5$ h) $-15 + 6$ i) $-15 + 14$

Sol.: a) 4; b) 3; c) 8; d) (–5); e) (–3); f) (–2); g) (–7); h) (–9); i) (–1).

8. Remove parentheses and calculate:

a) $-(-3) + (-7) - (-2) + (-9)$ b) $3 - 7 - (-2) + 9$

c) $15 + (-21) - 20 - (-15)$ d) $(-52) - 24$

e) $(-72) + 80 - 8$ f) $2 + (-3) + (-5) + (-4) + 6$

g) $(-3) + 6 + (-4) + 4 + (-6)$ h) $(-1) - 6 + (-7) - (-5) - 21 + (-12)$

Sol.: a) (-11); b) 7; c) (-11); d) (-76); e) 0; f) (-4); g) (-3); h) (-42).

9. Work out:

a) $-30 + 8 - (-5) + 1 - 5 - (-3) + (-7)$ b) $-4 + (-2 + 1) + 5 - [3 - (1-2) + 4] + 1 - 2$

c) $-19 + (-4) - (-8) + (-13) - (-12) + 4 - 57$

d) $3 - [-2 + 1 - (4 - 5 - 7)] - 2 + [-3 - (5 - 6 - 1) + 2]$

e) $-8 + (-2) - (-10) - 2 + 5$ f) $(3 - 8) + (-5 - 2) - (-9 + 1) - (7 - 5)$

g) $-[12 + (-3)] - (-4) - 5 + 6 - (-4)$

Sol.: a) (-25); b) (-9); c) (-69); d) (-5); e) 3; f) (-6); g) 0.

10. Work out:

a) $5 + [\,2 - ((\,4 + 5 - 3\,) + 6\,] - 1 - (\,3 + 5\,)$

b) $-4\,(\,4 - 5 + 2\,) - 3 - \{\,1 - [\,6 + (\,-3 - 1\,) - (\,-2 + 4\,)\,] + 3 - 4\,\}$

c) $10 - [\,-2 + (\,-3 - 4 - 1\,) + 1 - (\,-4 - 2 + 3 - 1\,) - 4\,]$

d) $(\,-6 + 4\,) - \{\,4 - [\,3 - (\,8 + 9 - 2\,) - 7\,] - 35 + (\,4 + 8 - 15\,)\,\}$

e) $-6 - \{\,-4 - [\,-3 - (\,1 - 6\,) + 5\,] - 8\,\} - 9$

f) $-3 + \{\,-5 - [\,-6 + (\,4 - 3\,) - (\,1 - 2\,)\,] - 5\,\}$

g) $-(\,9 - 15 + 2\,) + \{\,-6 + [\,4 - 1 + (\,12 - 9\,) + 7\,]\,\} - 3$

Sol.: a) (- 2); b) (-4); c) 19;   d) 13;   e) 4;   f) (-9);   g) 8.

11. Solve these operations:

a) $(-4) \cdot (+2) \cdot (-6)$ b) $(+8) \cdot (-3) \cdot (-4)$ c) $(-2) \cdot (-3) \cdot (-4)$

d) $(+20) : (+2) : (-5)$ e) $(-32) : (-4) : (-8)$ f) $(-80) : (-20) : (-4)$

Sol.: a) 48; b) 96; c) (-24); d) (-2); e) (-1); f) (- 1).

12. Work out:

a) $(+21) \cdot (+2) : (-14)$ b) $(+5) : (-5) \cdot (-4)$ c) $(+2) \cdot (+9) : (-3)$

d) $[(-2) \cdot (+7)] : (-14) \cdot (+3)$ e) $(+36) : [(-9) : (+3)] \cdot (+5)$ f) $(+36) : (-9) : (+2) \cdot (+5)$

Sol.: a) (-3); b) 4; c) (-6); d) 3; e) (-60); f) (-40).

13. How many years did Archimedes live, who was born in 287 BC and died in 212 BC? And Newton, who was born in 1642 and died in 1727? Draw each life in a Time Line.

Sol.: 75 and 85 years.

14. What temperature difference does a worker support when passing from a vegetables conservation room, at 4 °C, to the frozen product zone, at minus 18 degrees Celsius?

Sol.: (- 22) °C   ((-18) – 4 = (-22) °C).

15. When ascending in the atmosphere, air temperature decreases at a rate of 9 ° C per 300 m. If the temperature at sea level at a given point is 0 ° C, what is the height at which plane is flying if the air temperature is (-81) °C? **Sol.:** 2700 meters.

16. In a fish conservation ultra-refrigerator, temperature drops 3 ° C every 5 minutes. If we introduce a box of fishes at 9 ° C, a) how long will it take to reach -27 ° C? b) What will its temperature be after two hours and a half? **Sol.:** a) 1 hour;   b) (-81) °C.

17. Write as an only power:

a) $2^3 \cdot 2^4$ b) $3^5 \cdot 3^3 \cdot 3^4$ c) $(-7)^2 \cdot (-7)^5$ d) $4^5 : 4^2$ e) $7^6 : 7^3$

f) $x^9 \cdot x^4$ g) $(5^2)^3$ h) $((-2)^2)^3$ i) $(-2)^9 : (-2)^5$ j) $(-9)^7 : (-9)^4$

        Sol.: a) 2^7; b) 3^{12}; c) $(-7)^7$; d) 4^3; e) 7^3; f) x^{13}; g) 5^6; h) $(-2)^6 = 2^6$; i) $(-2)^4 = 2^4$; j) $(-9)^3$.

18. Write as an only power:

a) $5^4 \cdot 3^4$ b) $2^2 \cdot 4^2$ c) $3^4 \cdot 4^4 \cdot 5^4$ d) $x^n \cdot y^n$ e) $(-5)^4 \cdot (-3)^4$

f) $8^3 : 4^3$ g) $9^5 : 3^5$ h) $x^n : y^n$ i) $(-10)^4 : 2^4$ j) $(-6)^3 : (-3)^3$

        Sol.: a) 15^4; b) 8^2; c) 60^4; d) $(x \cdot y)^n$; e) 15^4; f) 2^3; g) 3^5; h) $(x : y)^n$; i) $(-5)^4 = 5^4$; j) 2^3.

19. Substitute the symbol (Δ) by a number:

a) $2^\Delta \cdot 2^{11} = 2^{20}$ b) $(-4)^6 : (-4)^\Delta = (-4)^3$ c) $(-6)^3 \times (-6)^5 = (-6)^\Delta$

d) $[(-1)7]^\Delta = (-1)^{21}$ e) $(3^{11})^2 = 3^\Delta$

        Sol.: a) 9;   b) 3;   c) 8;   d) 3; e) 22.

20. Write as an only power:

a) $4^3 \cdot 4^4 \cdot 4$ b) $(5^6)^3$ c) $\dfrac{7^6}{7^4}$ d) $\dfrac{15^3}{3^3}$ e) $2^{10} \cdot 5^{10}$ f) $\dfrac{12^5}{3^5 \cdot 4^5}$

g) $(a^6 \cdot a^3)^2 : (a^2 \cdot a^4)^3$ h) $(6^2)^3 \cdot 3^5 \cdot (2^7 : 2^2)$

        Sol.: a) 4^8;   b) 5^{18};   c) 7^2;   d) 5^3;   e) 10^{10};   f) 1;   g) 1;   h) 6^{11}.

21. In my city, there are 6 hostels. Each hostel has 6 rooms. In each room, there are 6 beds. Price per night at these hostels is 6 €/night. How much money will they earn one night in which all they are complete? Write it as a power and calculate its value.   **Sol.:** $6^4 = 1296$ euros.

22. Calculate by using the laws of powers:

a) $2^3 \cdot 5^4$ b) $(6^5 : 2^4) : 3^5$ c) $\left(\dfrac{2}{3}\right)^6 \cdot \left(\dfrac{3}{4}\right)^3$ d) $2^8 \left(\dfrac{5}{2}\right)^4$ e) $\dfrac{20^6}{2^6}$

f) $\dfrac{20^6}{2^5}$ g) $(3^3)^2 : 3^5$ h) $(2^5)^3 \cdot [(5^3)^4 : 2^3]$

        Sol.: a) 5000;   b) 2;   c) 1/27;   d) 10^4;   e) 10^6;   f) 2,000,000;   g) 3;   h) 10^{12}.

23. Write as an only power:

$2^{50} \cdot 2^{15} : 2^{60} =$

$(-3)^{115} \cdot (-3)^{172} : (-3)^{284} =$

$(5^{12})^3 \cdot 5^{31} : (5^5)^{13} =$

$(7^{15} \cdot 7^5)^3 : (7^{29})^2 =$

$(2^4)^{45} \cdot (2^3)^{30} : (2^2)^{134} =$

$11^{23} \cdot 11^{10} : 11^{31} =$

$(-6)^{80} \cdot (-6)^{15} : (-6)^{92} =$

$(2^{10})^3 \cdot 2^{18} : (2^4)^{11} =$

$(13^{18} \cdot 13^7)^4 : (13^{33})^3 =$

$(3^6)^4 \cdot (3^3)^{10} : (3^2)^{25} =$

Sol.: 2^6 ; $(-3)^3$; 5^2 ; 7^2 ; 2^2 ; 11^2 ; $(-6)^3$; 2^4; 13; 3^4.

24. Write as an only power:

$6^4 \cdot (6^5 : 6^2) =$

$\dfrac{2^{16} \cdot 2^{20}}{2^{50} : 2^{18}} =$

$\dfrac{11^{27} : 11^{21}}{11^{35} : 11^{30}} =$

$\dfrac{(2^5)^3 \cdot 2^6}{2^{20} : (2^3)^4} =$

$\dfrac{(6^7)^2 \cdot (6^3)^4}{6^{40} : (6^8)^4} =$

$(4^5 : 4^3)^6 =$

Sol.: 6^7 ; 2^4 ; 11 ; 2^{13} ; 6^{18} ; 4^{12}.

25. Calculate:

a) $(3-7)^2 : 8 - 2 \cdot 4 + 2 \cdot (6 + 4 \cdot 3)$

c) $5 + (-3)^3 + 64 : 4^2 - [6 + (6-2) \cdot 3] : 9$

b) $(-2)^3 \cdot [20 + (-2) \cdot 5]^2 + (-3 + 7)^2$

Sol.: a) 30; b) 816; c) (-20).

26. Work out the following roots. A possible answer might be *"It does not exist"*.

a) $\sqrt{64}$ b) $\sqrt[3]{-8}$ c) $\sqrt[4]{81}$ d) $\sqrt[5]{32}$ e) $\sqrt{-121}$

f) $\sqrt[3]{27}$ g) $\sqrt[4]{-16}$ h) $\sqrt[5]{-32}$

Sol.: a) 8 and (-8); b) (-2); c) 3 and (-3); d) 2; e) It does not exist; f) 3; g) It does not exist; h) (-2).

27. a) A squared room has an area of 144 m². How long is its side?

b) What is the volume of a cube having an edge of 5 cm?

c) If the side of a square is 14 hm, what is, in m², its area?

d) Calculate the edge of a cube whose volume is 216 m³.

Unit 3.- Decimal numbers

1. Decimal numbers. Reading decimal numbers

Decimal numbers, or simply, *decimals,* such as 3.762, are used in situations in which we look for more precision than whole numbers provide.

As with whole numbers, a digit in a decimal number has a value which depends on the place of the digit. In our decimal system, digits can be placed to the left and right of a decimal **point**, to indicate numbers greater than one or less than one, respectively. The decimal point helps us to keep track of where the "ones" place is. It's placed just to the right of the ones place. As we move right from the decimal point, each number place is divided by 10.

Note that adding extra zeros to the right of the last decimal digit does not change the value of the decimal number. This means that 123.456 = ...0000123.4560000.........

Place (underlined)	Name of Position
1.234567	Ones (units) position
1.234567	Tenths
1.234567	Hundredths
1.234567	Thousandths
1.234567	Ten thousandths
1.234567	Hundred Thousandths
1.234567	Millionths

	Whole-number part				Decimal point	Fractional part				
Ten thousands	Thousands	Hundreds	Tens	Ones		Tenths	Hundredths	Thousandths	Ten-thousandths	Hundred-thousandths
10,000	1000	100	10	1		$\frac{1}{10}$	$\frac{1}{100}$	$\frac{1}{1000}$	$\frac{1}{10,000}$	$\frac{1}{100,000}$
10^4	10^3	10^2	10^1	1		$\frac{1}{10^1}$	$\frac{1}{10^2}$	$\frac{1}{10^3}$	$\frac{1}{10^4}$	$\frac{1}{10^5}$

How do we write the name of a decimal number?

To write the name for a decimal number

1. Write the name for the whole number to the left of the decimal point.

2. Write the word *and* for the decimal point.

3. Write the whole number name for the number to the right of the decimal point.

4. Write the place value for the digit farthest to the right.

If the decimal has only zero or no digit to the left of the decimal point, omit steps 1 and 2.

Example 1:

0.83	is read as	Eighty-three hundredths
0.0027	is read as	Twenty-seven ten-thousandths
556.43	is read as	Five hundred fifty-six and forty-three hundredths

---

To write the place value name for a decimal

1. Write the whole number (number before the word *and*).

2. Write a decimal point for the word *and..*

3. Ignoring the place value name, write the name for the number following the word *and*. Insert zeros if necessary, between the decimal point and the digits following it to ensure that the place on the far right has the correct (given) place value.

Example 2:

- Two hundred thirty-seven and fifty-eight hundredths   is written   237.58

- Seven hundred twenty-three thousandths   is written   0.723

---

1. Write in words the following decimals:

a) 0.26	b) 0.82	c) 0.267	d) 0.943	e) 7.002	f) 4.3007
g) 11.92	h) 32.03	i) 0.805	j) 8.05	k) 80.05	l) 8.0 05
m) 61.0203	n) 45.0094	ñ) 90.003	o) 900.030	p) 21.456	q) 0.77
r) 0.0089	s) 5.7254				

2. Write the place value name.

a) Forty-two hundredths   b) Sixty-nine hundredths

c) Four hundred nine thousandths   d) Five hundred nineteen thousandths

e) Nine and fifty-nine thousandths   f) Sixteen and six hundredths

g) Three hundred eight ten-thousandths   h) Twelve ten-thousandths

How much would you say it costs??

It costs, aproximately, **2.5 €**

To round a decimal number to a given place value
1. Draw an arrow under the given place value. (After enough practice, you will be able to round mentally and will not need the arrow).
2. If the digit to the right of the arrow is 5, 6, 7, 8 or 9, add 1 to the digit above the arrow; that is, round to the larger number.
3. If the digit to the right of the arrow is 0, 1, 2, 3 or 4, keep the digit above the arrow; that is, round to the smaller number.
4. Write whatever zeros are necessary after the arrow so that the number above the arrow has the same place value as the original.

Example 3:

a) Round 0.7539 to the nearest hundredth.

$0.7539 \approx 0.75$ The digit to the right of the round-off place is 3, so round down.

b) Round 7843.9 to the nearest thousand.

7843.9 ≈ 8000   Three zeros must be written after the 8 to keep it in the
↑   thousands place.

c) Round 537.7 to the nearest unit.

537.7 ≈ 538   The number to the right of the round-off place is 7, so we
↑   round up by adding 537 + 1 + 538.

Exercises

3. Round to the nearest hundredth:

    a) 67.4856   b) 27.6372   c) 548.7235   d) 375.7545

4. Build a chart and, in it, round to the nearest tenth, hundredth and thousandth:

    a) 35.7834   b) 61.9639   c) 86.3278   d) 212.7364

    e) 0.91486   f) 0.8049

5. Round to the nearest dollar:

    a) $72.49   b) $38.51   c) $7821.51   d) $8467.80

---

3. Comparing decimals

Symbols < (less or lower than), > (greater or bigger than) and = (equals) are used to show how the size of one number compares to another.

To list a set of decimals from smallest to largest

1. Make sure that all numbers have the same number of decimal places to the right of the decimal point by writing zeros to the right of the last digit when necessary.

2. Write the numbers in order as if they were whole numbers.

3. Remove the extra zeros.

Example 4: List these numbers from smallest to largest: 4.502 4.511 4.6005 4.50102.

Solution: As you can see, the number that has more decimal digits is 4.50102, with 5 of them. So, we add zeros to the others, so that they all have 5 decimal digits → 4.50200; 4.51100; 4.60050; 4.50102.

If they were whole numbers, they would be → 450200; 451100; 460050; 450102.

Now, we order them as whole numbers → 450102 < 450200 < 451100 < 460050.

Finally, we substitute them by originals (removing added zeros) →

4.50102 < 4.502 < 4.511 < 4.6005.

6. Arrange these numbers in order of size, smallest first:

6.21,   6.023,   6.4,   6.04,   2.71,   9.4

7. Insert the symbols < or > between these pairs of numbers

a) 1.2__0.62   b) 1.23__1.3   c) 4.008__4.03

d) 0.24__0.204   e) 0.509__0.6   f) 1.582__1.59

8. Arrange these numbers in order of size, smallest first:

a) 6.1; 4.22; 4.02; 6.11; 3.99; 3.9   b) 5.602; 5.611; 5.6005; 5.60102

9. Arrange these numbers in order of size, smallest first:

a) 0.75; 0.57; 0.507; 0.705   b) 0.102; 0.05; 0.105; 0.501; 0.251

10. Insert a decimal number in each pair:

a) 4.5 and 4.6   b) 7.24 and 7.242   c) 2.3 and 2.4   d) 3.35 and 3.36

e) 5.23 and 5.24   f) 5.39 and 5.4   g) 3 and 3.1   h) 6.03 and 6.04

11. Is the statement true or false?

a) 0.38 < 0.3   b) 0.49 < 0.50   c) 10.48 > 10.84   d) 7.78 < 7.87

56

4. Fractions and decimals

Decimal expressions of rational numbers

As a fraction is the indicated quotient of two numbers. we can calculate these quotients only by dividing numerator and denominator.

For example. if we divide numerator by denominator of the fractions $\dfrac{221}{4}$. $\dfrac{25}{3}$ and $\dfrac{59}{6}$. we will

obtain their *decimal expressions*: $\dfrac{221}{4} = 55.25$ $\quad\quad$ $\dfrac{25}{3} = 8.3333........$ $\quad\quad$ $\dfrac{59}{6} = 9.833333.....$

As you can see. when dividing numerator by denominator will happen on of these situations:

· Division finishes because it is exact. as happens in the first fraction. In this case. we obtain a **limited** decimal expression (it has a limited number of decimal digits). Its decimal expression is said to be an *exact or terminating decimal number*.

· Division does not finish because it is not exact. as happens in second and third fractions. In these cases. we obtain a **non-limited** decimal expression (its decimal digits do not finish). Their decimal expressions are named *recurring* or *repeating decimal numbers* Second one is named *pure recurring decimal number*. and third one is a mixed recurring decimal number.

Repeating part is named *period* and it is indicated with an arch on it. For example:

$$2.333........ = 2.\overline{3} \quad\quad\quad\quad 1.8333........ = 1.8\overline{3}$$

So. decimal numbers can be classified into:

- **Exact** or **terminating** decimal: It finishes. so you can write down all its digits.

    For example: 0.125

- **Recurring** or **repeating** decimal: It does not finish. it goes on forever. but <u>some of the digits are</u> <u>repeated over and over again.</u>

    For example: $5.6666...... = 5.\overline{6}$ $14.353535....... = 14.\overline{35}$

        $87.42222...... = 87.4\overline{2}$

    Between recurring decimals. we can distinguish:

    - **Pure recurring decimals**: Recurring decimals in which period starts just after the decimal point.

        For example: $5.6666...... = 5.\overline{6}$ $14.353535....... = 14.\overline{35}$

    - **Mixed recurring decimals:** Recurring decimals in which period does not start just after The decimal point.   For example: $87.42222...... = 87.4\overline{2}$

- **Irrational** numbers: Decimals that cannot be written as fractions. They do not finish and they do not have any repeating part. For example: $\sqrt{2} = 1.4142135.........$ $\pi = 3.141592654..... $ $e = 2.718........$

Decimal fractions

Fractions whose decimal expression is a terminating decimal are named **decimal fractions.**

There is a very simple rule to identify decimal fractions without calculating their decimal expression:

A fraction is a *decimal fraction* if its denominator has as prime factors. only **2's** and/or **5's**.

12. Express in its decimal form the following fractions and indicate their type of decimal number:

a) $\dfrac{9}{2}$ b) $\dfrac{12}{7}$ c) $\dfrac{4}{3}$ d) $\dfrac{17}{100}$ e) $\dfrac{1}{6}$ f) $\dfrac{7}{495}$ g) $\dfrac{25}{99}$

13. Copy the following fractions in your notebook and round decimal fractions. You do not need to calculate them.

a) $\dfrac{19}{52}$ b) $\dfrac{41}{18}$ c) $\dfrac{37}{75}$ d) $\dfrac{17}{125}$ e) $\dfrac{1}{40}$ f) $\dfrac{7}{300}$ g) $\dfrac{8}{99}$

Rational expression of decimal numbers

All decimal numbers can be expressed as a fraction.

In following example we show the process to find out these fractions.

· **Exact decimal:** Express as a fraction $x = 2.34$

Solution: We must build a fraction whose

- Numerator: Decimal number without decimal point.

- Denominator: A "1" followed by as many "0" as decimal digits decimal number has.

In our case: $2.34 = \dfrac{234}{100}$ If possible. we must simplify this fraction to give it in its simplest

form $2.34 = \dfrac{224}{100} = \dfrac{112}{50} = \boxed{\dfrac{56}{25}}$

· **Recurring decimal number:** Express as a fraction $x = 14.\overline{35} = 14.3535355\ldots\ldots$

Solution: We must apply the following process. consisting on multiplying it by 1. 10. 100. 1000. ….. till finding two decimals having *identical decimal parts*. so that they can be subtracted. to cancel their decimal parts and converting them into integers.

$$100 \cdot x = 1435.353535....$$
$$10 \cdot x = 143.535353.....$$
$$\underline{x = 14.353535........}$$

$$100 \cdot x = 1435.353535....$$
$$\cancel{10 \cdot x = 143.535353.....}$$
$$\underline{x = 14.353535........}$$

$$99 \cdot x = 1421 \implies x = \dfrac{1421}{99}$$

As you can see. x and $100 \cdot x$ have the same decimal part. So. we subtract them:

Example 5: Express as a fraction: a) $8.\overline{5}$ b) $25.4\overline{52}$

Solution:

a)

$$10 \cdot x = 85.555.......$$
$$\underline{x = 8.5555......}$$

$$10 \cdot x = 85.555.......$$
$$\underline{x = 8.5555......}$$

$$9 \cdot x = 77 \implies x = \dfrac{77}{9}$$

As you can see. both decimals have the same decimal part. So. we subtract them:

59

b)

$$1000 \cdot x = 25452.5252......$$
$$100 \cdot x = 2545.25252......$$
$$10 \cdot x = 254.525252....$$
$$x = 25.4525252....$$

As you can see. $1000 \cdot x$ and $10 \cdot x$ have the same decimal part. So. we subtract them:

$$1000 \cdot x = 25452.5252......$$
$$\cancel{100 \cdot x = 2545.25252......}$$
$$\cancel{10 \cdot x = 254.525252....}$$
$$x = 25.4525252....$$

$$990 \cdot x = 25198 \implies x = \frac{25198}{990}$$

Exercises

14. Write each of the following *exact* decimals as a fraction:

a) 23.45   b) 5.101   **Sol.:** a) 469/20 ; b) 5101/1000.

15. Write each of the following *pure recurring* decimals as a fraction:

a) $24.\overline{3}$ b) $0.\overline{678}$ c) $32.\overline{76}$ d) $1.\overline{876}$ e) $26.\overline{7}$

16. Write each of the following *mixed recurring* decimals as a fraction:

a) $2.7\overline{5}$ b) $34.6\overline{41}$ c) $2.09\overline{5}$ d) $0.2\overline{3}$ e) $5.12\overline{04}$

f) $103.25\overline{6}$ g) $0.0\overline{23}$

17. Write each of the following decimals as a fraction:

a) $1,0\overline{4} =$ b) $7,\overline{123} =$ c) $8,17\overline{4} =$ d) $1,7\overline{6} =$ e) $2,5\overline{19} =$

18. Write each of the following decimals as a fraction:

a) $23.\overline{45}$ b) $76.8\overline{1}$ c) 398.12   d) 5.6   e) $28.\overline{7}$

f) $261.1\overline{23}$ g) $0.0\overline{4}$ h) 23.05

60

5. Adding and subtracting decimal numbers

To add or subtract decimals, line up the decimal points and then follow the rules for adding or subtracting whole numbers, placing the decimal point in the same column as above.

When one number has more decimal places than another, use zeros to give them the same number of decimal places.

To add decimal numbers

1. Write in columns with the decimal points aligned. Insert extra zeros to help align the place values.
2. Add the decimals as if they were whole numbers.
3. Align the decimal point in the sum with those above.

Example 5:   Add $43.67 + 2.3$ Subtract $57.8 - 8.06$

$$
\begin{array}{r}
a)\quad 43.67 \\
+\ 2.30 \\
\hline
45.97
\end{array}
\qquad
\begin{array}{r}
b)\quad 57.80 \\
-\ 8.06 \\
\hline
49.74
\end{array}
$$

19. Add:

a) $0.7 + 0.7$ b) $0.6 + 0.5$ c) $3.7 + 2.2$ d) $7.6 + 2.9$

e) $1.6 + 5.5 + 8.7$ f) $6.7 + 2.3 + 4.6$ g) $34.8 + 5.29$ h) $22.9 + 7.67$

20. The sum of 6.7, 10.56, 5.993, and 45.72 has _____ decimal places.

21. Add:

a) $2.337 + 0.672 + 4.056$ b) $9.445 + 5.772 + 0.822$

c) $0.0017 + 1.007 + 7 + 1.071$ d) $1.0304 + 1.4003 + 1.34 + 0.403$

22. Subtract:

a) 0.831 - 0.462   b) 0.067 - 0.049   c) 33.456 - 29.457

d) 7.598 - 4.7732   e) 327.58 - 245.674   f) 506.5065 - 341.341

23. On a vacation trip, Manuel stopped for gas four times. The first time, he bought 19.2 gallons. At the second station he bought 21.9 gallons, and at the third, he bought 20.4 gallons. At the last stop, he bought 23.7 gallons. How much gas did he buy on the trip?

24. What is the total cost of a cart of groceries that contains bread for $3.09, bananas for $1.49, cheese for $2.50, cereal for $4.39, coffee for $7.99, and meat for $9.27? Round the result to the tenths.

25. The table shows the lengths of railway tunnels in various countries.

World's Longest Railway Tunnels

Tunnel	Length (km)	Country
Seikan	53.91	Japan
English Channel Tunnel	49.95	UK–France
Dai-shimizu	22.53	Japan

a) How much longer is the longest tunnel than the second longest tunnel?

b) What is the total length of the Japanese tunnels?

26. In 2004, the average interest rate on a 30-year home mortgage was 6.159%. In 2009, the average interest rate was 4.759%. What was the drop in interest rate?

27. How high from the ground level is the top of the tree shown below? Round to the nearest foot.

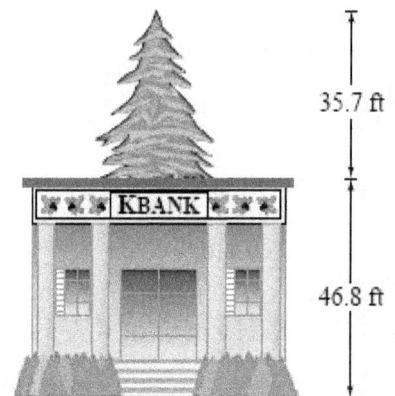

35.7 ft

46.8 ft

28. What is the center-to-center distance, *A*, between the holes in the diagram?

6. Multiplying decimal numbers

Multiplying decimals is just like multiplying whole numbers. The only extra step is to decide how many digits to leave to the right of the decimal point. To do that, add the numbers of digits to the right of the decimal point in both factors.

To multiply decimals

1. Multiply the numbers as if they were whole numbers.

2. Locate the decimal point by counting the number of decimal places (to the right of the decimal point) in both factors. The total of these two counts is the number of decimal places the product must have.

3. If necessary, zeros are inserted to the left of the numeral so there are enough decimal places.

Example 6: Multiply 23.56 x 34.1

Solution:

$$
\begin{array}{r}
23.56 \\
\times\ 3.1 \\
\hline
2356 \\
7068 \\
\hline
73036 \\
\end{array}
$$

Decimal digits: 3

29. Calculate:

      a) 5.6 x 6.9   b) 12.37 x 76.78   c) 4.66 · 4.7   d) 0.345 · (32.4 – 4.67)

30. Find the product of 9.73 and 6.8.

31. Multiply 7.9 times 0.0004.

32. Multiply 32 x 0.846 and round the product to the nearest tenth.

33. The table shows calories expended for some physical activities.

Calorie Expenditure for Selected Physical Activities

Activity	Step Aerobics	Running (7 min/mile)	Cycling (10 mph)	Walking (4.5 mph)
Calories per pound of body weight per minute	0.070	0.102	0.050	0.045

  a) Vanessa weighs 145 lb and does 75 min of step aerobics per week. How many calories per week does she burn per week?

  b) Steve weighs 187 lb and runs 25 min five times at a 7 min/mi pace. How many calories does he burn per week?

34. An order of 43 bars of steel is delivered to a machine shop. Each bar is 17.325 ft long. Find the total linear feet of steel in the order.

17.325 ft

43 bars

35. In the 2009 World Championships in Athletics, Shelly-Ann Fraser of Jamaica won the 100 m with a time of 10.73 sec. If she could continue that rate, what would her time for the 400 m be?

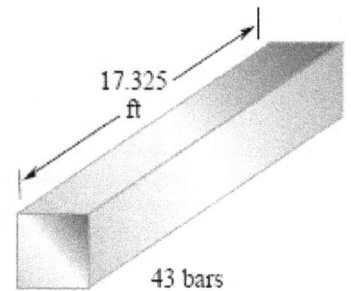

Exercises

36. In the 2009 World Championships in Athletics, Usain Bolt set a new world record for 200 m, with a time of 19.19 sec. Assuming he ran the first 100 m in his record time of 9.58 sec, how long did the second 100 m take him?

7. Multiplying and dividing by a power of ten

Sometimes, e.g. when dividing two decimals, it is necessary to multiply a decimal number by a power of ten. But it will be more interesting the inverse process, converting a decimal number containing a large quantity of zeros into the product of a decimal and a power of ten.

To **multiply** a decimal by a power of ten, we must move the decimal point to the right. The number of places to move is shown by the exponent in the power of 10.

To **divide** a decimal by a power of ten, we must move the decimal point to the left. The number of places to move is shown by the exponent in the power of 10.

In both cases, we can add zeros if necessary.

* IMPORTANT: If the exponent of the power of ten is negative, the movements have an inverse direction.

Example 7: Calculate:
  a) $34.7561 \cdot 10^2$
  b) $12345.8 \cdot 10^{-3}$
  c) $12.5878 : 10^2$
  d) $3.4 : 10^{-3}$.

Solution:

  a) $34.7561 \cdot 10^2 = 3475.61$
  b) $12345.8 \cdot 10^{-3} = 12.3458$

  c) $12.5878 : 10^2 = 0.125878$
  d) $3.4 : 10^{-3} = 3400$

65

37. Calculate the following products:

a) $34.7 \cdot 10^2$ b) $0.892 \cdot 10$ c) $83 \cdot 10^{-3}$ d) $4.5 \cdot 10^4$ e) $23 \cdot 10^{-2}$

f) $45.2 \cdot 10^{-3}$ g) $0.98 \cdot 10^{-1}$

38. Calculate the following quotients:

a) $34.7 : 10^2$ b) $0.892 : 10^{-1}$ c) $83 : 10^3$ d) $4502.1 : 10^4$ e) $7.41 : 10^6$

f) $231.6 : 0'1$ g) $583 : 10^{-4}$ h) $2.14 : 10^3$ i) $3901.3 : 10^{-3}$

39. Calculate the following products and quotients:

a) $34.12 \cdot 10^3$ b) $2120.8 \cdot 10^{-5}$ c) $28.1 : 10^2$ d) $432 \cdot 10^{-4}$

e) $5.786 \cdot 10^5$ f) $2.45 : 10^3$ g) $8.123 : 10^{-2}$ h) $456.1 : 10^{-1}$

8. Dividing decimal numbers

8.1. Dividing whole numbers, with decimals

Sometimes, when dividing two whole numbers, if the division is not exact, we can calculate the whole quotient and remainder, or we can also continue the whole division adding zeros to the right of the dividend until we get the amount of decimal digits required.

Example 8: Divide 235:6 until the hundredth:

```
235|00 |6
 55|     39.16
  1|0
   |40
    4
```

40. Calculate with two decimal digits:

a) 56 : 7   b) 7634 : 34   c) 679 : 32   d) 9783 : 127

8.2. Dividing decimals by decimals

To divide by a decimal,

1. Multiply that decimal by a power of 10 great enough to obtain a whole number.

2. Multiply the dividend by that same power of 10.

3. Now the problem becomes one involving division by a whole number instead of division by a decimal.

Example 9: Divide 0.35789 : 0.12 until the hundredths:

Solution: First of all, we must multiply $0.12 \times 10^2 = 12$,   and also $0.35789 \times 10^2 = 35.789$

$$0.35789 \times 100 \quad | \quad 0.12 \times 100$$

35|789 |12
11|7₈   2.982
⁰0|98
   |2 9
     5 //

41. Calculate with three decimal digits:

a) 56.7 : 2.34   b) 1432.3 : 0.42   c) 12.34 : 3.5   d) 1 : 1.2

42. 17 tickets cost £ 21.25. If they all cost the same, find the cost of one ticket.

43. A bottle contains 0.9 liters of lemonade. How many glasses with a capacity of 0.15 liters, can be filled from it?

44. A milkman is carrying a crate which contains 12 bottles and weighs 11.5 kg. If the empty crate weighs 0.7 kg, what is the weight of each bottle of milk?

Review exercises

1. Write the word name.

    a) 6.12   b) 0.843   c) 15.058   d) 0.0000027

2. Write the place value name.

    a) Twenty-one and five hundredths   b) Four hundred nine ten-thousandths

    c) Four hundred and four hundredths   d) One hundred twenty-five and forty-five thousandths.

3. Round the numbers to the nearest tenth, hundredth, and thousandth.

	Tenth	Hundredth	Thousandth
9. 34.7648			
10. 7.8736			
11. 0.467215			

4. The display on Mary's calculator shows 91.457919 as the result of a division exercise. If she is to round the answer to the nearest thousandth, what answer does she report?

5. List the set of decimals from smallest to largest.

    a) 0.95, 0.89, 1.01   b) 0.09, 0.093, 0.0899

    c) 7.017, 7.022, 0.717, 7.108   d) 34.023, 34.103, 34.0204, 34.0239

6. Is the statement true or false?

    a) 6.1774 > 6.1780   b) 87.0309 < 87.0319

7. Find the sum of 3.405, 8.12, 0.0098, 0.3456, 11.3, and 24.9345.

8. Find the difference of 56.7083 and 21.6249.

9. Find the perimeter of the following figure:

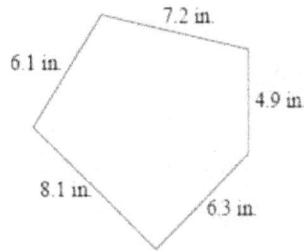

10. Multiply: 0.074 x 2.004. Round to the nearest thousandth.

11. Multiply: 0.0098 x 42.7. Round to the nearest hundredth.

12. Find the area of the rectangle.

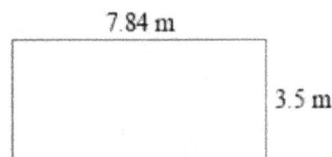

13. Multiply or divide:

a) $13.765 \cdot 10^3$ b) $7.023 \cdot 10^6$ c) $0.7321 \cdot 10^5$ d) $9.503 \cdot 10^2$

14. Multiply or divide:

a) $7 \cdot 10^7$ b) $8.13 \cdot 10^{-6}$ c) $6.41 \cdot 10^{-2}$ d) $3.505 \cdot 10^3$

15. Divide, rounding to the nearest tenth:

a) 0.3 : 0.0111   b) 75 : 40.5   c) 56.7 : 0.32

16. Carol drove 375.9 km on 12.8 gallons liters of fuel. How many kilometers does she drive on each liter? Calculate with two decimal digits.

17. The Metropolis Police Department reported the following number of robberies for the week:

Monday: 12   Tuesday: 21   Wednesday: 5   Thursday: 18

Friday: 46   Saturday: 67   Sunday: 17.

To the nearest tenth, what is the average number of robberies reported per day?

18. Perform the indicated operations:

      a) $0.65 + 4.29 - 2.71 + 3.04$ b) $13.8 : 0.12 \cdot 4.03$

      c) $(6.7)^2 - 4.4 \cdot 2.93$ d) $2.4 \cdot (3.02 + 0.456) - (9.231 + 0.4)$

      e) $100.15 : 100 - 3.995 \cdot 0.05$

19. Ellen earns £137.40 per week and after 4 weeks she gets an extra payment of £24.75. She spends £354.60 in this period. How much has she saved?

20. Susan cooked a cake and used 1.35 kg of flour, 0,37 kg of sugar, 3 eggs that weigh 80 g each and 240 g of milk. Which is the weight of the mixture?

21. I bought 7 mugs and a glass and paid 58 €. The price of the glass was 4.45€ . How much did each mug cost?

22. Express as a fraction:

    a)   2.41   b)   0.75   c) $3.\overline{5}$ d) $1.0\overline{3}$

23. Express as a fraction and check that the result of these operations is an integer:

    a) $2.333\ldots + 4.666\ldots$ b) $6.171717\ldots + 3.828282\ldots$

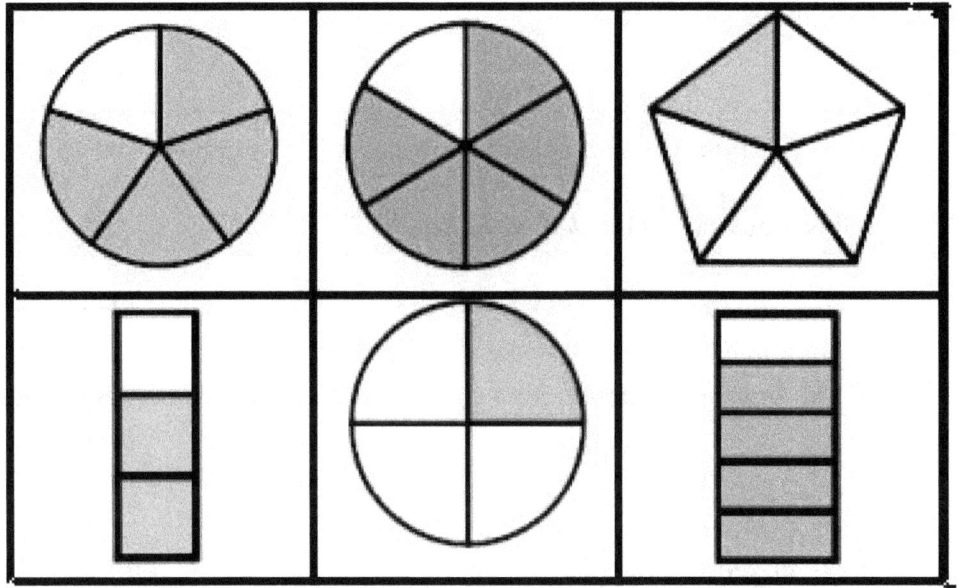

Unit 4.- Fractions

1. Concept of fraction

A *fraction* is a part of a whole.

If we cut a cake into two equal pieces, and then eat one of them, we say that we have eaten $\frac{1}{2}$ *(half)* a cake.

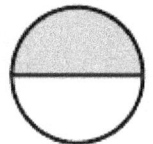

If we cut a cake into five equal pieces, then eat three of them, we say that we have eaten $\frac{3}{5}$ *(three fifths)* of a cake.

$\frac{1}{2}$ and $\frac{3}{5}$ are examples of fractions (parts of a whole).

Some definitions

- A *fraction,* (for example, $\frac{3}{5}$) is a name for a number. The upper numeral is the **numerator**. The lower numeral is the **denominator**.

- A *proper fraction* is one in which the numerator is less than the denominator. For example, $\frac{5}{12}$.

- An *improper fraction* is one in which numerator is not less than denominator. For example, $\frac{17}{12}$ or $\frac{12}{12}$.

- A **mixed number** is the sum of a whole number and a fraction with the addition sign omitted. For example, $\frac{17}{5} = 3 + \frac{2}{5}$ and it can be written as a mixed number as $3\frac{2}{5}$.

The fraction part is usually a proper fraction.

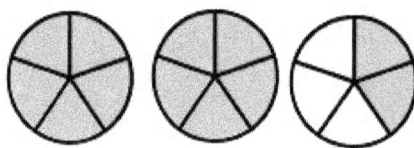

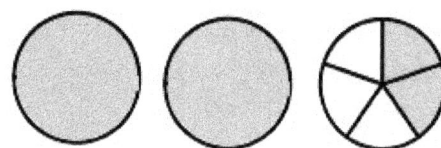

This can be seen graphically.

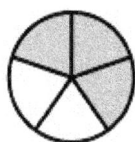

$\dfrac{3}{5}$	$\dfrac{12}{5}$	$2\dfrac{2}{5}$
Proper fraction	Improper fraction	Mixed number

2. Reading fractions

The method to name the denominator of a fraction depends on the denominator:

- If denominator ≤ 10, we use the ordinals to name it.

    For example, we read $\frac{2}{5}$ as *two fifths,* we read $\frac{9}{6}$ as *nine sixths.*

72

There are two exceptions, for denominators 2 and 4: $\frac{1}{2}$ is a half and $\frac{3}{4}$ is three quarters.

- For denominators larger than 10 we say "over" and do not use ordinal, so we read:

$\frac{4}{11}$ is *four over eleven* $\frac{5}{20}$ is *five over twenty* $\frac{6}{17}$ is *six over seventeen.*

1. What part of the shapes are shaded? Write in numbers and letters:

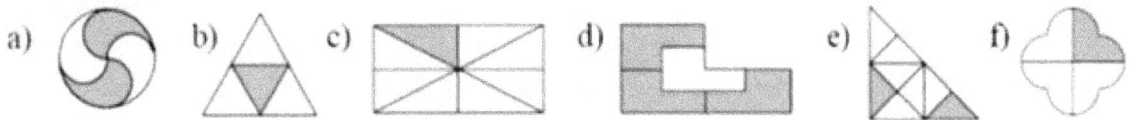

a)   b)   c)   d)   e)   f)

2. Colour the given fraction of each shape:

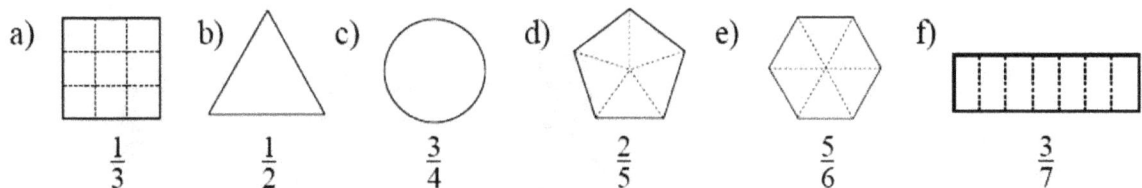

a)   b)   c)   d)   e)   f)

$\frac{1}{3}$ $\frac{1}{2}$ $\frac{3}{4}$ $\frac{2}{5}$ $\frac{5}{6}$ $\frac{3}{7}$

3. In a martial arts course there are 13 women and 15 men. What fraction of people do men represent and what women?

4. In a safari section of the zoo there are 7 zebras, 3 lions, 5 hippos and 1 giraffe.

   a) How many animals are there together?

   b) What fraction of the animals are zebras?

   c) What fraction of the animals are lions?

5. Write down the way you read these fractions:

   a) $\frac{5}{8}$ b) $\frac{12}{9}$ c) $\frac{5}{14}$.

6. What fraction of 1 hour is:

   a) 5 minutes: b) 15 minutes: c) 40 minutes:

Imagine you are eating pizza with some friends. One of your friends cuts a pizza into 3 pieces and he eats 1 of them.

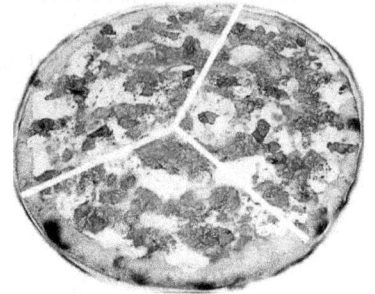

After that, you cut another pizza into 6 pieces and eat 2 of them. Who has eaten more pizza?

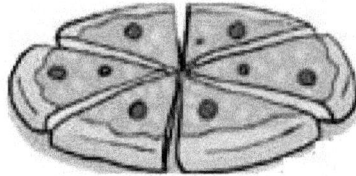

Notice you have eaten the same. These two fractions are *EQUIVALENT*

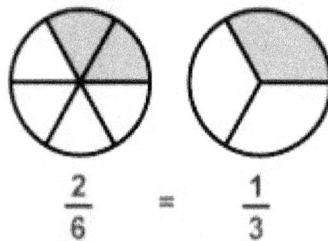

$$\frac{2}{6} = \frac{1}{3}$$

Equivalent fractions are different fractions that name the same amount.

For example:

$$\frac{2}{6} = \frac{1}{3} \qquad \frac{1}{3} = \frac{4}{12}$$

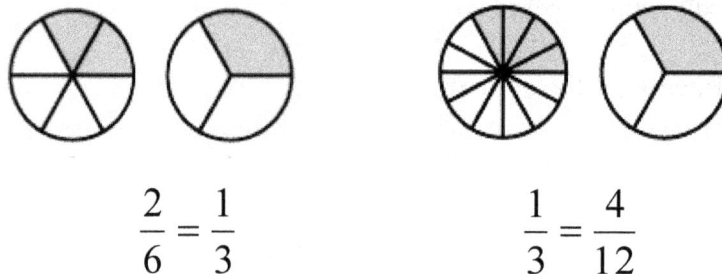

Equivalent fractions are fractions that represent equal values, even though they look different. Starting with any fractions you like, you can make up a list of equivalent fractions by simply multiplying or dividing both the numerator and the denominator by the same number each time.

7. What part of the shapes are shaded? Are they equivalent?

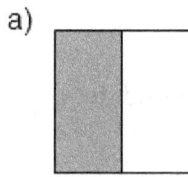

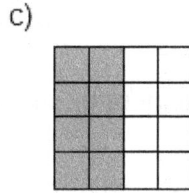

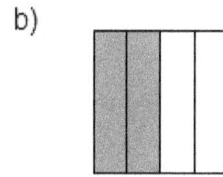

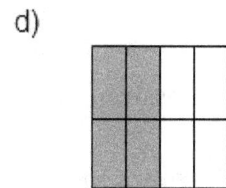

a)   c)   b)   d)

8. Copy the following figures in your notebook and colour the figures on the right to make them equivalent to the ones on the left. Write their fractions.

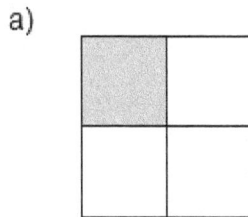

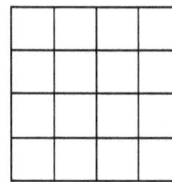

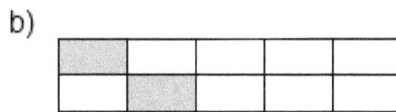

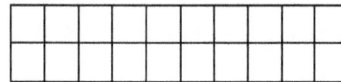

a)

b)

9. Write four equivalent fractions to the following:

a) $\dfrac{4}{6} =$

b) $\dfrac{15}{18} =$

c) $\dfrac{14}{21} =$

10. Write equivalent fractions, having the denominator given:

a) $\dfrac{2}{3} = \dfrac{}{12} = \dfrac{}{6}$

b) $\dfrac{12}{15} = \dfrac{}{5} = \dfrac{}{30}$

c) $\dfrac{7}{2} = \dfrac{}{24} = \dfrac{}{14}$

We can test if two fractions are equivalent by cross-multiplying their numerators and denominators. This is also called taking the cross-product.

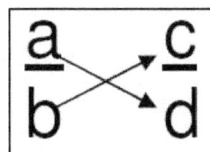

$$\dfrac{a}{b} \times \dfrac{c}{d}$$

Example 1: Check if $\dfrac{12}{20}$ and $\dfrac{24}{40}$ are equivalent fractions.

Solution:

1st cross-product is the product of the first numerator and the second denominator: $12 \cdot 40 = 480$.

2nd cross-product is the product of the second numerator and the first denominator: $24 \cdot 20 = 480$.

Since the cross-products are the same, the fractions are equivalent.

11. Check if the each pair of fractions are equivalent:

a) $\dfrac{9}{15}$ and $\dfrac{6}{10}$

b) $\dfrac{6}{18}$ and $\dfrac{3}{9}$

c) $\dfrac{20}{21}$ and $\dfrac{4}{14}$

12. Find out the value of x so that each pair of fractions are equivalent:

a) $\dfrac{4}{6} = \dfrac{x}{9}$

b) $\dfrac{4}{10} = \dfrac{6}{x}$

c) $\dfrac{x}{21} = \dfrac{6}{9}$

d) $\dfrac{12}{20} = \dfrac{21}{x}$

13. Indicate if the following equalities are true or false:

a) $\dfrac{-2}{6} = \dfrac{1}{3}$

b) $\dfrac{9}{2} = \dfrac{27}{6}$

c) $\dfrac{-4}{12} = \dfrac{-5}{15}$

d) $\dfrac{-2}{5} = \dfrac{-8}{20}$

e) $\dfrac{6}{2} = \dfrac{12}{5}$

14. In the list there are equivalent fractions. Find them:

$$\dfrac{2}{6}, \ \dfrac{6}{8}, \ \dfrac{10}{6}, \ \dfrac{3}{9}, \ \dfrac{12}{16}, \ \dfrac{5}{15}, \ \dfrac{5}{3}, \ \dfrac{9}{12}$$

15. Complete to make the following equalities true:

a) $\dfrac{6}{21} = \dfrac{10}{}$

b) $\dfrac{}{4} = \dfrac{3}{12}$

c) $\dfrac{3}{2} = \dfrac{}{12}$

d) $\dfrac{7}{2} = \dfrac{14}{}$

4. Simplifying fractions. Simplest form of a fraction

Simplifying a fraction is the process of renaming it by using a smaller numerator and denominator.

When a fraction is *completely simplified* (its numerator and denominator have no common factors other than 1), it is s named the ***simplest form*** of the fraction.

There are two available methods to simplify a fraction.

Method 1: Its useful for small numbers.

We must divide the numerator and denominator by a common factor. Keep dividing until there are no more common factors.

Example 2: Find the simplest form of $\dfrac{12}{30}$:

Solution: $\dfrac{12}{30} = \dfrac{12 \div 2}{30 \div 2} = \dfrac{6 \div 3}{15 \div 3} = \boxed{\dfrac{2}{5}}$

NOTE: If numerator and denominator have as last digits zeros, we can simplify all the common zeros above and below. For example, find the simplest form of $\dfrac{6000}{4500}$:

Solution: $\dfrac{6000}{4500} = \dfrac{60 \otimes \otimes}{45 \otimes \otimes} = \dfrac{60}{45} = \dfrac{60 \div 5}{45 \div 5} = \dfrac{12 \div 3}{9 \div 3} = \boxed{\dfrac{4}{3}}$

Exercises

16. Express these fractions in the simplest form. (Some may already be in the simplest form)

a) $\dfrac{5}{10}$ b) $\dfrac{30}{36}$ c) $\dfrac{18}{27}$ d) $\dfrac{48}{84}$ e) $\dfrac{45}{66}$ f) $\dfrac{64}{88}$

17. Express, in the simplest form, which fraction corresponds to these situations:

a) In a bag of 90 pens, 15 are blue

b) The number of girls and boys in our class

c) There are 90 pupils of the 270 who come by bus to the school.

Method 2: General method.

We must find the prime factorization of both numerator and denominator. Then, eliminate all common factors, other than 1, in the numerator and the denominator.

Example 3: Find the simplest form of $\dfrac{120}{100}$:

Solution: First, find the prime factorizations:

$$\left.\begin{array}{l} 120 = 2\cdot 2\cdot 2\cdot 3\cdot 5 \\ 100 = 2\cdot 2\cdot 5\cdot 5 \end{array}\right\}$$ Then, simplify the common factors *(Do it by yourself, with your pencil!!!)*

$$\left.\begin{array}{l} 120 = 2\cdot 2\cdot 2\cdot 3\cdot 5 \\ 100 = 2\cdot 2\cdot 5\cdot 5 \end{array}\right\} \Rightarrow \frac{120}{100} = \frac{2\cdot 2\cdot 2\cdot 3\cdot 5}{2\cdot 2\cdot 5\cdot 5} = \frac{2\cdot 3}{5} = \boxed{\frac{6}{5}}$$

NOTE: If numerator and denominator have as last digits zeros, you should simplify them at the beginning.

For example, find the simplest form of $\dfrac{7500}{5000}$:

Solution: $\dfrac{7500}{5000} = \dfrac{75\otimes\otimes}{50\otimes\otimes} = \dfrac{75}{50} = \left\{\begin{array}{l} 75 = 3\cdot 5\cdot 5 \\ 50 = 2\cdot 5\cdot 5 \end{array}\right\} = \dfrac{3\cdot 5\cdot 5}{2\cdot 5\cdot 5} = \boxed{\dfrac{3}{2}}$

Exercises

18. Simplify the following fractions by using the general method:

a) $\dfrac{66}{55}$ b) $\dfrac{360}{540}$ c) $\dfrac{700}{4900}$ d) $\dfrac{11\cdot 7\cdot 2^2\cdot 3}{11\cdot 2^3\cdot 3}$

19. Simplify the following fractions by using the general method:

a) $\dfrac{126}{210}$ b) $\dfrac{126}{252}$ c) $\dfrac{2^3\cdot 3^2\cdot 5\cdot 2^4}{5^2\cdot 2^5\cdot 3^2}$

Imagine you are in a party. You have eaten $\frac{2}{5}$ of a pizza and your friend Andrés has eaten $\frac{4}{7}$ of another pizza. Who has eaten more?

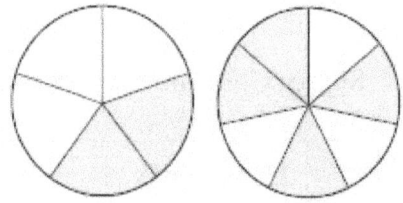

It is hard to answer this question just by looking at the fractions.

But, would it be easier if you had had $\frac{2}{5}$ of a pizza and Andrés $\frac{3}{5}$?

What is the difference?   *THEY HAVE THE SAME DENOMINATORS!!!*

So, first of all, we must transform our original fractions into two new fractions, equivalent to original ones and having the same denominator, the Lowest Common Multiple of both denominators.

This is going to be necessary also when adding or subtracting fractions, so, pay attention to this transformation into Lowest Common Denominator.

Example 4: Compare $\frac{2}{5}$ and $\frac{4}{7}$:

Solution:

Step 1: Find out the LCM(5, 7) = 35

Step 2: Transform both original fractions into two new fractions having 35 as denominator. Remember you can multiply numerator and denominator by the same number:

$\frac{2}{5} = \frac{}{35} \Rightarrow$ To transform 5 into 35, we have multiplied by 7, so, we must also multiply the numerator

by 7. So, as, $2 \cdot 7 = 14$, $\frac{2}{5} = \frac{14}{35}$

$$\frac{4}{7} = \frac{}{35} \Rightarrow$$ To transform 7 into 35, we have multiplied by 5, so, we also multiply the numerator

by 5. So, as, $4\cdot 5 = 20$, $\frac{4}{7} = \frac{20}{35}$

<u>Step 3:</u> Compare the new fractions: As $\frac{14}{35} < \frac{20}{35} \Rightarrow\Rightarrow\Rightarrow \frac{2}{5} < \frac{4}{7}$

20. Which is bigger?

a) $\frac{3}{5} or \frac{7}{15}$ b) $\frac{3}{7} or \frac{6}{21}$ c) $\frac{10}{15} or \frac{4}{6}$ d) $\frac{1}{3} or \frac{3}{100}$

21. Put these fractions in ascending order of size: a) $\frac{8}{3}, \frac{6}{4} and \frac{12}{5}$ b) $\frac{3}{10}, \frac{13}{20} and \frac{2}{3}$

22. Insert the symbol <, > or =:

a) $\frac{2}{15}$ $\frac{3}{10}$ b) $\frac{7}{30}$ $\frac{5}{12}$ c) $\frac{3}{20}$ $\frac{7}{12}$ d) $\frac{3}{20}$ $\frac{2}{15}$ $\frac{7}{10}$

6. Addition and subtraction of fractions

What is the sum of $\frac{1}{5}$ and $\frac{2}{5}$?

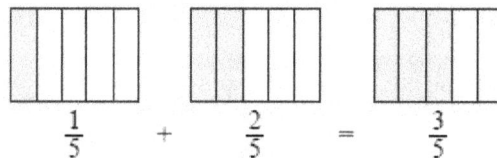

$$\frac{1}{5} \quad + \quad \frac{2}{5} \quad = \quad \frac{3}{5}$$

What is the difference of $\frac{3}{5}$ and $\frac{1}{5}$?

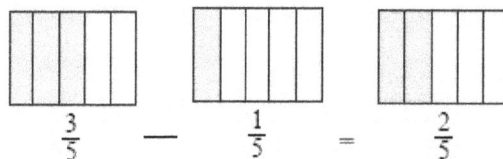

$$\frac{3}{5} \quad - \quad \frac{1}{5} \quad = \quad \frac{2}{5}$$

If the fractions have the same denominator the numerator of the sum is found by simply adding the numerators over the denominator. Their difference is the difference of the numerators over the denominator.

We do not add or subtract the denominators!!!

If the fractions have different denominators, it is a bit harder. Let's see it with an example:

Example 5: Calculate $\dfrac{2}{5} + \dfrac{4}{7}$:

<u>Solution:</u>

<u>Step 1:</u> Find out the LCM(5, 7) = 35

<u>Step 2:</u> Transform both original fractions into two new fractions having 35 as denominator.

Remember you can multiply numerator and denominator by the same number:

$\dfrac{2}{5} = \dfrac{}{35} \Rightarrow$ To transform 5 into 35, we have multiplied by 7, so, we must also multiply the

numerator by 7. So, as, $2 \cdot 7 = 14$, $\dfrac{2}{5} = \dfrac{14}{35}$

$\dfrac{4}{7} = \dfrac{}{35} \Rightarrow$ To transform 7 into 35, we have multiplied by 5, so, we must also multiply the

numerator by 5. So, as, $4 \cdot 5 = 20$, $\dfrac{4}{7} = \dfrac{20}{35}$

<u>Step 3:</u> As new fractions have the same denominator, they can be added: $\dfrac{2}{5} + \dfrac{4}{7} = \dfrac{14}{35} + \dfrac{20}{35} = \boxed{\dfrac{34}{35}}$

In this case, the final fraction cannot be simplified.

---

23. Calculate the following sums and subtractions of fractions having the same denominator:

a) $\dfrac{5}{6} - \dfrac{20}{6} + \dfrac{7}{6}$

b) $\dfrac{7}{4} + \dfrac{13}{4} - \dfrac{6}{4}$

c) $\dfrac{9}{15} + \dfrac{9}{15} + \dfrac{7}{15}$

d) $\dfrac{3}{8} + \dfrac{5}{8} - \dfrac{18}{8}$

e) $\dfrac{15}{9} - \dfrac{5}{9} - \dfrac{7}{9}$

f) $\dfrac{7}{12} - \dfrac{25}{12} + \dfrac{16}{12}$

24. Calculate the following sums and subtractions of fractions having different denominator:

a) $\dfrac{1}{6} - \dfrac{7}{9} + \dfrac{2}{3}$

b) $\dfrac{4}{15} + \dfrac{1}{6} - \dfrac{7}{10}$

c) $2 - \dfrac{3}{4} - \dfrac{5}{6}$

d) $\dfrac{7}{10} - \dfrac{1}{3} - \dfrac{7}{15}$

e) $\dfrac{3}{4} + \dfrac{5}{6} - \dfrac{2}{3}$

f) $\dfrac{2}{5} - \dfrac{2}{8} - \dfrac{7}{20}$

g) $\dfrac{5}{4} + \dfrac{7}{6} - \dfrac{9}{2}$

h) $\dfrac{5}{2} - \dfrac{7}{3} + \dfrac{1}{6}$

Exercises

25. Calculate:

a) $\dfrac{11}{10}-\left(\dfrac{2}{5}-\dfrac{1}{20}\right)$ b) $\left(\dfrac{7}{9}+\dfrac{5}{6}\right)-\left(\dfrac{4}{15}-\dfrac{1}{18}\right)$ c) $\left(\dfrac{6}{7}+\dfrac{2}{9}\right)-\dfrac{3}{4}$ d) $\left(\dfrac{5}{4}-\dfrac{2}{5}\right)+\left(\dfrac{7}{9}-\dfrac{3}{12}\right)$

26. Convert the mixed number to improper fractions, operate and convert the results to mixed numbers.

a) $5\dfrac{1}{3}-\left(3\dfrac{2}{3}+\dfrac{4}{5}\right)+4\dfrac{2}{15}$ b) $3\dfrac{1}{3}+\dfrac{13}{2}-\left(\dfrac{1}{2}-\dfrac{1}{4}\right)+3\dfrac{5}{12}$ c) $7\dfrac{3}{9}+2\dfrac{1}{3}-\dfrac{12}{27}$

27. Jane used ½ of a piece of ribbon and her sister used 1/3 of it. What fraction of the ribbon was used?

28. Joe painted 2/5 of a fence and Bill painted ½ of it. What fraction of the fence did the boys paint?

29. Mrs Bell made 40 cookies. Her son ate 1/5 of them. How many cookies did he eat?

30. Harban was given £ 15 allowance each week. He spent 3/5 of it. What fraction did he save? How much did he save in pounds?

31. Mrs Holland spends ¼ of her money in the market and 1/3 **of the remainder** in a shop. What fraction of her money is left?

32. Joan earns £ 1800 a month. She spends 3/8 of her salary every month. She gives her parents 2/5 of the remainder and saves the rest. How much money does she save every month?

7. Product of fractions

To *multiply* two or more *fractions*, multiply the numerators and multiply the denominators, separately. Then simplify if necessary. $\dfrac{a}{b}\cdot\dfrac{c}{d}=\dfrac{a\cdot c}{b\cdot d}$

Example 6: Calculate: a) $\dfrac{3}{8} \cdot \dfrac{5}{2}$ b) $\dfrac{5}{7} \cdot \dfrac{2}{3}$

Solution:

a) $\dfrac{3}{8} \cdot \dfrac{5}{2} = \dfrac{3 \cdot 5}{8 \cdot 2} = \dfrac{15}{16}$ b) $\dfrac{5}{7} \cdot \dfrac{2}{3} = \dfrac{5 \cdot 2}{7 \cdot 3} = \dfrac{10}{21}$

Important: If possible, you must simplify the resulting fraction. And you can do this at the end, as you already know, but you can also simplify before multiplying. Let's see it with an example:

Example 7: Calculate $\dfrac{6}{8} \cdot \dfrac{10}{15}$

Solution: - Simplifying at the end) $\dfrac{6}{8} \cdot \dfrac{10}{15} = \dfrac{6 \cdot 10}{8 \cdot 15} = \dfrac{60}{120} = \dfrac{6}{12} = \dfrac{6 \div 2}{12 \div 2} = \dfrac{3 \div 3}{6 \div 3} = \boxed{\dfrac{1}{2}}$

     - Simplifying before) $\dfrac{6}{8} \cdot \dfrac{10}{15} = \dfrac{6 \cdot 10}{8 \cdot 15} = \dfrac{(2 \cdot 3) \cdot (2 \cdot 5)}{(2 \cdot 2 \cdot 2) \cdot (3 \cdot 5)} = \dfrac{2 \cdot 3 \cdot 2 \cdot 5}{2 \cdot 2 \cdot 2 \cdot 3 \cdot 5} = \boxed{\dfrac{1}{2}}$

Note: Sometimes, it is possible to simplify even before:

Example 8: Calculate $\dfrac{6}{10} \cdot \dfrac{10}{15}$ Solution: $\dfrac{6}{10} \cdot \dfrac{10}{15} = \dfrac{6 \cdot 10}{10 \cdot 15} = \dfrac{6}{15} = \dfrac{2 \cdot 3}{3 \cdot 5} = \boxed{\dfrac{2}{5}}$

Exercises

33. Calculate the following products of fractions, and give the result in their simplest forms:

a) $\left(-\dfrac{8}{15}\right) \cdot \dfrac{6}{4}$ b) $4 \cdot \dfrac{3}{10}$ c) $\dfrac{2}{5} \cdot \left(-\dfrac{10}{6}\right)$ d) $\dfrac{6}{15} \cdot \dfrac{10}{3}$ e) $\left(-\dfrac{7}{8}\right) \cdot \left(-\dfrac{6}{5}\right)$

f) $\dfrac{5}{18} \cdot 12$ g) $\dfrac{6}{15} \cdot \left(-\dfrac{5}{3}\right)$ h) $\dfrac{8}{6} \cdot \dfrac{9}{20}$ i) $\left(-\dfrac{10}{9}\right) \cdot \dfrac{6}{5}$ j) $\left(-\dfrac{7}{2}\right) \cdot \left(-\dfrac{6}{14}\right)$

k) $\dfrac{7}{8} \cdot \dfrac{9}{14}$ l) $\dfrac{12}{7} \cdot \left(-\dfrac{15}{6}\right) \cdot \dfrac{9}{10}$ m) $\dfrac{5}{7} \cdot \dfrac{3}{5} \cdot \dfrac{7}{3}$ n) $\dfrac{3}{4} \cdot \dfrac{2}{3} \cdot \dfrac{7}{8}$ ñ) $\dfrac{2}{9} \cdot \dfrac{3}{7} \cdot \dfrac{14}{15}$

34. Calculate the following products of fractions, and give the result in their simplest forms:

a) $\left(-\dfrac{26}{3}\right) \cdot \left(-\dfrac{1}{2}\right) \cdot \left(\dfrac{-9}{39}\right)$ b) $\left(1 - \dfrac{4}{7}\right) \cdot \left(\dfrac{1}{3} + \dfrac{1}{2}\right)$ c) $\left(\dfrac{2}{7} - 2\right) \cdot \left(1 - \dfrac{5}{4} - \dfrac{25}{12}\right)$

35. Two fifths of my garden is lawn and 1/3 of the rest is a flower bed. How much of my garden is left for a vegetable plot?

Calculating the fraction of a whole

Example 9: What is $\dfrac{2}{5}$ of 35?

Solution: $\dfrac{1}{5}$ of 35 = 35 ÷ 5 = 7 ⇒⇒⇒ $\dfrac{2}{5}$ of 35 = 2 x 7 = 14.

Example 10: How much is the $\dfrac{5}{8}$ of 32 kg?

Solution: $\dfrac{1}{8}$ of 32 = 32 ÷ 8 = 4 ⇒⇒⇒ $\dfrac{5}{8}$ of 32 = 5 x 4 = 20.

---

36. Work out:

a) $\dfrac{1}{8}$ of 72   b) $\dfrac{4}{9}$ of 63   c) $\dfrac{5}{6}$ of 36   d) $\dfrac{3}{4}$ of 36

37. A chain store closed $\dfrac{2}{15}$ of its 345 shops. How many shops were closed?

38. Mum bought 1200 g of grapes. John ate 2 fifths of them, Betty ate 1 quarter of them and dad ate the rest. What amount of grapes did each of them eat?

39. Joan earns £ 1800 a month. She spends $\dfrac{3}{8}$ of her salary every month. She gives her parents $\dfrac{2}{5}$ of the remainder and saves the rest. How much money does she save every month?

40. Mrs Holland spends $\dfrac{1}{4}$ of her money in the market and $\dfrac{1}{3}$ of the remainder in a shop. What fraction of her money is left?

41. Harban was given £ 15 allowance each week. He spent $\dfrac{3}{5}$ of it. What fraction did he save? How much did he save in pounds?

Exercises

8. Quotient of two fractions

To *divide* two fractions, turn the second fraction *UPSIDE DOWN* and then multiply them.

$$\frac{a}{b} : \frac{c}{d} = \frac{a}{b} \cdot \frac{d}{c} = \frac{a \cdot d}{b \cdot c}$$

Example 11: Calculate: $\dfrac{3}{8} : \dfrac{5}{3}$ <u>Solution:</u> $\dfrac{3}{8} : \dfrac{5}{3} = \dfrac{3}{8} \cdot \dfrac{3}{5} = \dfrac{3 \cdot 3}{8 \cdot 5} = \dfrac{9}{40}$

Important: If possible, you must simplify the resulting fraction. And you can do this at the end, as you already know, but you can also simplify before multiplying. Let´s see it with an example:

Example 12: Calculate $\dfrac{6}{8} : \dfrac{15}{10}$

<u>Solution:</u>

Simplifying at the end) $\dfrac{6}{8} : \dfrac{15}{10} = \dfrac{6}{8} \cdot \dfrac{10}{15} = \dfrac{6 \cdot 10}{8 \cdot 15} = \dfrac{60}{120} = \dfrac{6}{12} = \dfrac{6 \div 2}{12 \div 2} = \dfrac{3 \div 3}{6 \div 3} = \boxed{\dfrac{1}{2}}$

Simplifying before) $\dfrac{6}{8} : \dfrac{15}{10} = \dfrac{6}{8} \cdot \dfrac{10}{15} = \dfrac{6 \cdot 10}{8 \cdot 15} = \dfrac{(2 \cdot 3) \cdot (2 \cdot 5)}{(2 \cdot 2 \cdot 2) \cdot (3 \cdot 5)} = \dfrac{2 \cdot 3 \cdot 2 \cdot 5}{2 \cdot 2 \cdot 2 \cdot 3 \cdot 5} = \boxed{\dfrac{1}{2}}$

Note: Sometimes, it is possible to simplify even before:

Example 13: Calculate $\dfrac{6}{10} : \dfrac{15}{10}$

<u>Solution:</u> $\dfrac{6}{10} : \dfrac{15}{10} = \dfrac{6}{10} \cdot \dfrac{10}{15} = \dfrac{6 \cdot 10}{10 \cdot 15} = \dfrac{6}{15} = \dfrac{2 \cdot 3}{3 \cdot 5} = \boxed{\dfrac{2}{5}}$

Exercises	**42.** Calculate the following quotients fractions, and give the result in their simplest forms:

42. Calculate the following quotients fractions, and give the result in their simplest forms:

a) $\left(-\dfrac{4}{9}\right) : \left(-\dfrac{2}{6}\right)$ b) $8 : \dfrac{12}{5}$ c) $\dfrac{5}{6} : \dfrac{10}{4}$ d) $\left(-\dfrac{12}{5}\right) : 10$ e) $\left(-\dfrac{6}{15}\right) : \dfrac{9}{2}$

f) $\dfrac{10}{25} : \dfrac{4}{5}$ g) $\left(-\dfrac{10}{4}\right) : \dfrac{5}{3}$ h) $\dfrac{6}{15} : \left(-\dfrac{3}{8}\right)$ i) $\left(-\dfrac{2}{3}\right) : \left(-\dfrac{20}{25}\right)$ j) $\left(-\dfrac{9}{2}\right) : 6$

43. Complete:

a) $\dfrac{36}{25} : \dfrac{}{5} = \dfrac{4}{5}$

b) $\dfrac{}{3} : \dfrac{14}{6} = 1$

c) $\dfrac{3}{25} \cdot \dfrac{}{9} = \dfrac{1}{5}$

d) $\dfrac{2}{3} \cdot \dfrac{9}{} = \dfrac{3}{2}$

e) $\dfrac{10}{} : \dfrac{5}{14} = 4$

f) $\dfrac{12}{8} \cdot \dfrac{15}{9} \cdot \dfrac{7}{35} = \underline{}$

44. You have to walk 7/4 km to school. How far have you walked when you are halfway?

45. A recipe for 6 buns requires 3/2 kg of sugar. How much sugar is needed for 1 bun?

46. A plank of wood is 35/8 meters long. How many pieces of wood of length of 5/4 metres can be cut from the piece of wood?

47. Work out the following operations:

a) $\dfrac{3}{5} + \dfrac{7}{5} \cdot \dfrac{8}{7}$

b) $\dfrac{1}{6} + \dfrac{2}{3} \cdot \dfrac{6}{5} - \dfrac{2}{10} : \dfrac{1}{3} + \dfrac{2}{3}$

c) $\dfrac{2}{7} + \dfrac{2}{7} \cdot \dfrac{14}{3} - \dfrac{5}{3}$

d) $\dfrac{2}{13} \cdot \dfrac{26}{3} - \dfrac{2}{5} \cdot \dfrac{10}{3}$

e) $\dfrac{2}{5} - \dfrac{2}{5} \cdot \left(\dfrac{3}{4} - \dfrac{2}{6} \right) - \left(2 - \dfrac{1}{2} \right)$

9. Problems with fractions

48. In a factory a metal rod 49 meter long is divided into 7/5 meter pieces. How many pieces can be made from each rod?

49. A math's teacher spends 4/5 of his working time in school teaching his classes. If 2/9 of this teaching time is spent with his first year classes. What fraction of his working time does he spend teaching the first year?

50. A seven tons load of soil container is divided in 3/140 tons bags. How many bags can be filled?

51. Last night John spent ¾ of an hour on English homework and 5/6 of this time, on his maths. If he started his homework at 20:00 o'clock, did he finish it in time to watch The Simpsons, which started at 21:00 o'clock? Explain your answer.

52. Dolly opened a ¾ liter bottle of coke, James drank 1/3 litter and Carole 1/5 litter. How much coke was left for Dolly?

53. There are 300 passengers on a train. At a station, 3/5 of the passengers get off. How many people get off the train? How many people are left on the train?

54. Allan has 120€. He decides to save 2/5 of this and to spend 1/6 on books. How much does he save? How much does he spend on books? How much is left?

10. Powers of fractions

We have already studied powers. The same rules for powers of integers are also applied when base is a fraction. In the next table, you are shown some examples that should help you to understand and remember the rules of powers.

Laws
$\left(\dfrac{a}{b}\right)^m \cdot \left(\dfrac{a}{b}\right)^n = \left(\dfrac{a}{b}\right)^{(m+n)}$
$\left(\dfrac{a}{b}\right)^m : \left(\dfrac{a}{b}\right)^n = \left(\dfrac{a}{b}\right)^{(m-n)}$
$\left(\left(\dfrac{a}{b}\right)^m\right)^n = \left(\dfrac{a}{b}\right)^{m \cdot n}$
$\left(\dfrac{a}{b}\right)^1 = \dfrac{a}{b}$
$\left(\dfrac{a}{b}\right)^0 = 1$

Example: Simplify:

a) $\left(\dfrac{3}{5}\right)^5 \cdot \left(\dfrac{3}{5}\right)^6 \cdot \left(\dfrac{3}{5}\right)^{-8} = \left(\dfrac{3}{5}\right)^{(5+6+(-8))} = \left(\dfrac{3}{5}\right)^3$

b) $\left(-\dfrac{5}{7}\right)^5 : \left(-\dfrac{5}{7}\right)^2 = \left(-\dfrac{5}{7}\right)^{(5-2)} = \left(-\dfrac{5}{7}\right)^3$

c) $\left(-\dfrac{2}{3}\right)^5 : \left(-\dfrac{2}{3}\right)^{-3} = \left(-\dfrac{2}{3}\right)^{(5-(-3))} = \left(-\dfrac{2}{3}\right)^8$

d) $\left(-\dfrac{5}{4}\right)^1 = \left(-\dfrac{5}{4}\right)$

e) $\left(-\dfrac{21154}{31585}\right)^0 = 1$

55. Simplify:

a) $\left(\dfrac{5}{8}\right)^5 \cdot \left(\dfrac{5}{8}\right)^{-2}$ b) $\left(-\dfrac{2}{5}\right)^6 : \left(-\dfrac{2}{5}\right)^{-3}$ c) $\left(\dfrac{1}{7}\right)^5 \cdot \left(\dfrac{1}{7}\right)^{-5}$ d) $\left(-\dfrac{8}{7}\right)^{-1} \cdot \left(-\dfrac{8}{7}\right)^{-8}$ e) $\left[\left(\dfrac{2}{7}\right)^{-6}\right]^{-3}$

f) $\left[\left(\dfrac{7}{10}\right)^{-3}\right]^2$ g) $\left[\left(\left(\dfrac{2}{9}\right)^{-2}\right)^{-5}\right]^{-3}$ h) $\dfrac{\left[\left(\dfrac{5}{2}\right)^3 \cdot \left(\dfrac{5}{2}\right)^{-2}\right]^{-5}}{\left[\left(\dfrac{5}{2}\right)^{-5} : \left(\dfrac{5}{2}\right)^{-9}\right]^{-3}}$ i) $\dfrac{\left(\dfrac{2}{7}\right)^{-3} : \left(\dfrac{2}{7}\right)^{-8}}{\left(\dfrac{2}{7}\right)^2 \cdot \left(\dfrac{2}{7}\right)^{-8} : \left(\dfrac{2}{7}\right)}$

10.1. Power of a fraction

It might be useful to expand a power of a fraction, having a positive or a negative exponent:

$$\left(\dfrac{a}{b}\right)^n = \dfrac{a^n}{b^n} \qquad \left(\dfrac{a}{b}\right)^{-n} = \dfrac{b^n}{a^n}$$

When we have a ***negative exponent*** in a factor into numerator or denominator, we can change it UP/DOWN if we *change* its sign.

Example: Simplify:

a) $\dfrac{3 \cdot 2^{-4}}{2^3} = \dfrac{3}{2^3 \cdot 2^4} = \boxed{\dfrac{2}{2^7}}$

b) $\dfrac{5^5 \cdot 7^6 \cdot 5^{-5} \cdot 7^{-8}}{5^{-2} \cdot 7^3} = \dfrac{5^5 \cdot 7^6 \cdot 5^2}{5^5 \cdot 7^3 \cdot 7^8}$ Now, for each base, we leave it where it has a higher total exponent, and its exponent will be its total exponent there, minus its total exponent in the other part of the fraction:

$\dfrac{5^5 \cdot 7^6 \cdot 5^2}{5^5 \cdot 7^3 \cdot 7^8} = \dfrac{5^{(7-5)}}{7^{(11-6)}} = \boxed{\dfrac{5^2}{7^5}}$

c) $\dfrac{\left(3^2 \cdot 3^{-6} \cdot 5 \cdot 5^{-4}\right)^6}{\left(3 \cdot 5^3 \cdot 5^{-1}\right)^2}$ → If possible, simplify first into the parenthesis: $\dfrac{\left(3^{-4} \cdot 5^{-3}\right)^6}{\left(3 \cdot 5^2\right)^2}$

Then remove parenthesis by applying power of a power: $\dfrac{3^{-24} \cdot 5^{-18}}{3^2 \cdot 5^4}$

Now, change position of negative exponents by changing their signs:

$$\dfrac{1}{3^{24} \cdot 3^2 \cdot 5^{18} \cdot 5^4} = \boxed{\dfrac{1}{3^{26} \cdot 5^{22}}}$$

Exercises

55. Simplify:

a) $\dfrac{2^4 \cdot 2^{-3} \cdot 2^{-5} \cdot 2}{2^3 \cdot 2^5 \cdot 2^{-6}}$ b) $\dfrac{3 \cdot 2^4 \cdot 3^2 \cdot 2 \cdot 3^{-5}}{2^3 \cdot 2^2 \cdot 3^{-2} \cdot 2 \cdot 3^{-1}}$ c) $\dfrac{5 \cdot 2^{-3} \cdot 5^{-2} \cdot 2^6}{5^2 \cdot 2^{-4} \cdot 5^{-2} \cdot 2^3}$ d) $\dfrac{7 \cdot 5^3}{5^{-2} \cdot 14}$

e) $\dfrac{2^{-3} \cdot 3^2}{2^6 \cdot 5 \cdot 3^{-4}}$

Review exercises

1. Simplify the following fractions:

a) $\dfrac{66}{55}$ b) $\dfrac{360}{540}$ c) $\dfrac{700}{4900}$ d) $\dfrac{11 \cdot 7 \cdot 2^2 \cdot 3}{11 \cdot 2^3 \cdot 3}$ e) $\dfrac{4 \cdot 75}{20 \cdot 9}$

f) $\dfrac{30}{20}$ g) $\dfrac{50}{30}$ h) $\dfrac{126}{210}$ i) $\dfrac{126}{252}$ j) $\dfrac{2^3 \cdot 3^2 \cdot 5 \cdot 2^4}{5^2 \cdot 2^5 \cdot 3^2}$

2. Simplify the following fractions:

a) $\dfrac{4 \cdot 14}{21 \cdot 8}$ b) $\dfrac{4 \cdot 3 + 2}{4 \cdot 5}$ c) $\dfrac{15 \cdot 9}{3 \cdot 18}$ d) $\dfrac{36 \cdot 25}{30 \cdot 60}$ e) $\dfrac{6^2 \cdot 15^4}{10^4 \cdot 27^2}$

f) $\dfrac{12^2 \cdot 15^4}{10^4 \cdot 27^2}$ g) $\dfrac{7 \cdot 4 + 2}{7 \cdot 4 \cdot 3}$ h) $\dfrac{1 + 2 + 3}{1 + 2 + 3}$

3. List these fractions from lower size to bigger:

a) $\dfrac{3}{5}, \dfrac{2}{5}, \dfrac{1}{4}, \dfrac{1}{7}$ b) $\dfrac{2}{9}, \dfrac{3}{5}, \dfrac{6}{15}$ c) $\dfrac{6}{8}, \dfrac{5}{4}, \dfrac{5}{6}, \dfrac{10}{8}$ d) $\dfrac{4}{5}, \dfrac{7}{3}, \dfrac{9}{12}$

4. Calculate the following sums with fractions:

a) $\dfrac{2}{5} + \dfrac{7}{10}$

b) $\dfrac{6}{35} + \dfrac{3}{10} + \dfrac{1}{14}$

c) $\dfrac{3}{10} - \dfrac{5}{2} + \dfrac{2}{15}$

d) $\dfrac{4}{3} - \dfrac{5}{3} + \dfrac{2}{3}$

e) $\dfrac{1}{2} + \dfrac{5}{6} - \dfrac{7}{18}$

f) $\dfrac{4}{25} + \dfrac{1}{30} - \dfrac{2}{15}$

g) $-\dfrac{7}{4} - \dfrac{3}{12} + \dfrac{1}{6}$

h) $\dfrac{5}{27} + \dfrac{10}{54} - \dfrac{4}{9}$

i) $\dfrac{3}{7} - \dfrac{2}{7} + \dfrac{6}{7}$

j) $\dfrac{6}{3} - \dfrac{75}{5} + \dfrac{8}{2}$

k) $\dfrac{2}{3} + \dfrac{7}{9} - \dfrac{5}{6}$

l) $\dfrac{6}{3} + \dfrac{2}{5} - 3$

m) $8 - \dfrac{3}{5} + 2 - \dfrac{2}{5}$

n) $\dfrac{7}{2} + \dfrac{1}{5} - \dfrac{3}{2} + \dfrac{4}{5}$

ñ) $3 + \dfrac{5}{2}$

o) $\dfrac{2}{5} + \dfrac{1}{5}$

p) $\dfrac{3}{7} + \dfrac{2}{7} - \dfrac{1}{7}$

q) $5 - \dfrac{1}{2} + \dfrac{4}{5} - 7$

r) $\dfrac{4}{10} + \dfrac{9}{15} - \dfrac{8}{5}$

s) $\dfrac{5}{3} - \dfrac{2}{3} - \dfrac{4}{3} + \dfrac{1}{3}$

t) $\dfrac{3}{2} + \dfrac{5}{2} - \dfrac{9}{2} - \dfrac{7}{2} + \dfrac{1}{2}$

u) $\dfrac{1}{6} + \dfrac{5}{12} - \dfrac{7}{18}$

v) $\dfrac{2}{10} - \dfrac{5}{14} + \dfrac{3}{35} - \dfrac{8}{7}$

w) $\dfrac{12}{15} - \dfrac{17}{15} + \dfrac{20}{15}$

x) $\dfrac{23}{11} - \dfrac{31}{11} + \dfrac{13}{11} - \dfrac{8}{11} + \dfrac{25}{11}$

5. Calculate the following products of fractions, and give the result in their simplest forms:

a) $\dfrac{6}{10} \cdot \dfrac{5}{3}$

b) $\dfrac{8}{7} \cdot \dfrac{14}{24} \cdot \dfrac{15}{5}$

c) $\dfrac{2}{9} \cdot \dfrac{18}{6} \cdot \dfrac{3}{2}$

d) $\dfrac{42}{15} \cdot \dfrac{10}{98} \cdot \dfrac{14}{3}$

e) $\dfrac{56}{9} \cdot \dfrac{54}{25} \cdot \dfrac{5}{84}$

f) $3 \cdot \dfrac{5}{2}$

g) $7 \cdot \dfrac{3}{14}$

h) $4 \cdot \dfrac{1}{6} \cdot \dfrac{3}{2}$

i) $\dfrac{5}{3} \cdot 6 \cdot \dfrac{1}{20}$

j) $11 \cdot \dfrac{3}{22} \cdot \dfrac{1}{6}$

6. Calculate the following quotients of fractions, and give the result in their simplest forms:

a) $\dfrac{6}{10} : \dfrac{5}{3}$

b) $\left(\dfrac{8}{7} : \dfrac{24}{14} \right) : \dfrac{5}{15}$

c) $\left(\dfrac{2}{9} : \dfrac{6}{18} \right) : \dfrac{2}{3}$

d) $\left(\dfrac{42}{15} : \dfrac{98}{10} \right) : \dfrac{3}{14}$

e) $\left(\dfrac{56}{9} : \dfrac{25}{54} \right) : \dfrac{84}{5}$

f) $\dfrac{2}{5} : \dfrac{4}{3}$

7. Work out the following operations:

a) $\dfrac{3}{5} + \dfrac{3}{5} \cdot \dfrac{1}{3} - \dfrac{3}{10} : \dfrac{1}{4}$

b) $\left(\dfrac{2}{7} - 2 \right) \cdot \left(1 - \dfrac{5}{4} - \dfrac{25}{12} \right)$

c) $\left(\dfrac{3}{5} - \dfrac{1}{2} \right) : \dfrac{3}{10}$

8. In a magazine there are three adverts on the same page. Advert A uses ¼ of the page, advert B uses 1/8 and advert C uses 1/16 of the page.
   a) What fraction of the page do the three adverts use?
   b) An advert uses 3/6 of the page, if the cost of an advert is 12 € for each 1/32 of the page, how much does it cost?

9. A farmer owns 360 hectares of land. He plants potatoes on 3/10 of his land and beans on 1/6 of the remainder. How many hectares are planted with potatoes? How many hectares are planted with beans? How many hectares are left?

10. A journey is 120 miles. Richard has driven 3/5 of this distance and in a second stage 5/6 of the rest. How much farther does he have to drive to complete the journey?

11. At a sale shirts are sold by 3/5 of their original price and the sale price is 36 €. What was the original price?

12. Joe spends 3/8 of his salary on his own, gives 3/5 of the remainder to his parents and saves 450 €. What is his salary?

13. Calculate:

a) $\left(\dfrac{1}{2}\right)^3$ b) $\left(\dfrac{1}{3}\right)^2$ c) $\left(\dfrac{1}{5}\right)^4$ d) $\left(\dfrac{1}{10}\right)^6$

14. Calculate by the shortest way:

a) $\dfrac{6^2}{2^2}$ b) $\dfrac{18^3}{6^3}$ c) $\dfrac{9^4}{3^4}$

15. Simplify the following fractions. First, factorize numerator and denominator:

a) $\dfrac{20^2}{8^3}$ b) $\dfrac{70^2}{28^3}$ c) $\dfrac{12^4}{16^2}$ d) $\dfrac{14^3}{21}$ e) $\dfrac{25^2}{15^2}$ f) $\dfrac{15^3}{30^2}$

16. Simpliy:

a) $\dfrac{3^2 \cdot 5^6 \cdot 3^{-5} \cdot 5^{-2}}{3^4 \cdot 5^7}$ b) $\dfrac{2^5 \cdot 2^3 \cdot 2^{-2} \cdot 2^{-7}}{2^4 \cdot 2^{-3}}$ c) $\dfrac{3^{-5} \cdot 3^5 \cdot 3^{-2}}{3^4 \cdot 3^{-6}}$ d) $\dfrac{5^7 \cdot 5^{-3} \cdot 5}{5^{-8} \cdot 5}$ e) $\dfrac{7 \cdot 7^{-3} \cdot 7^2}{7^4 \cdot 7^{-1}}$

f) $\dfrac{3^3 \cdot 5^7 \cdot 3^{-1} \cdot 5}{5^{11} \cdot 3 \cdot 5^{-2}}$ g) $\dfrac{2^4 \cdot 2^{-3} \cdot 5 \cdot 7^3 \cdot 5^2}{7^4 \cdot 5^3 \cdot 2^{-2}}$ h) $\dfrac{3^6 \cdot 3^{-4} \cdot 5^3 \cdot 7^3}{5^4 \cdot 7^4 \cdot 3^{-2}}$ i) $\dfrac{5 \cdot 11^2 \cdot 5^{-3} \cdot 7}{11^3 \cdot 7^3 \cdot 5^{-2}}$ j) $\dfrac{\left(3^2 \cdot 3^{-2} \cdot 5 \cdot 5^{-4}\right)^6}{\left(3 \cdot 5^2 \cdot 5^{-1}\right)^2}$

k) $\dfrac{\left(5^3 \cdot 3 \cdot 11^2\right)^2}{\left(3^2 \cdot 5 \cdot 11^{-1}\right)^3}$

17. Calculate and simplify, expressing the results with positive exponents:

a. $\left(\dfrac{2}{3}\right)^6 \cdot \left(\dfrac{2}{3}\right)^{-4}$ b. $\left(\dfrac{3}{5}\right)^7 : \left(\dfrac{3}{5}\right)^{-2}$ c. $\left(\dfrac{1}{3}\right)^5 \cdot \left(\dfrac{1}{3}\right)^{-3}$ d. $\left(\dfrac{3}{4}\right)^{-2} \cdot \left(\dfrac{3}{4}\right)^{-5}$

e. $\left[\left(\dfrac{4}{5}\right)^3\right]^{-2}$ f. $\left[\left(\dfrac{2}{3}\right)^{-3}\right]^2$ g. $\left[\left(\dfrac{2}{3}\right)^{-3}\right]^{-2}$

Unit 5.- Proportionality and percentages

1. Introduction to proportionality

From the earliest times, humans have drawn maps to represent the geography of their surroundings. Some maps depict features encountered on a journey, like rivers and mountains.

The most useful maps incorporate the concept of scale, or proportion. Simply put, a scaled map accurately preserves relative distances. So if the distance from one city to another is twice the distance from the city to a river in real life, the distance between the cities is twice the distance from the city to a river on the map as well.

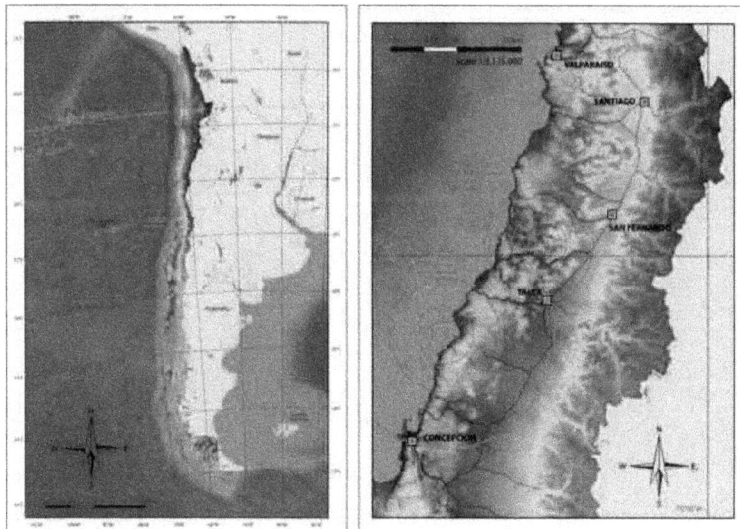

2. Direct and inverse proportionality

We say that there is a **direct proportionality** between two magnitudes if when we double one magnitude, the other also doubles, when we half the first, the second also halves.

We say that there is an **inverse proportionality** between two magnitudes if when increasing one magnitude, (double, triple...) the other decreases (half, third...), when decreasing one (half, third...), the other increases (double, triple...).

Example 1: Direct and inverse proportionality examples.

a) If a can of cola costs 40 cent, the cost of:
   - 2 cans is 80 cent
   - 5 cans is € 2.00 (200 cent).

   We can see that, for example, if we double the number of cans, we double the price. They are in **direct proportion**.

b) Look at the relationship that exists between the number of the members of a family and the days that one box of apples lasts them (suppose that all the people eat the same amount of apples at the same rate). Observe that the more people there are in the family the less time the box of fruit lasts, and the less people there are, the longer it lasts. These two magnitudes are in **inverse proportion**.

1. Classify the relationship of proportionality (direct, inverse or non-proportional) between the following pairs of magnitudes:

   a) The number of children attending a birthday party and the size of the piece of a cake that corresponds to each.

   b) The number of workers repairing a street and the number of days it takes them.

   c) Number of people attending a doctor for an hour and the time spent with each patient.

   d) The width of the shelf and the number of books (same type) you can put.

   e) Hours of operation of a machine and number of parts produced.

   f) The volume of a deposit and the time we need to fill it, using the same supplier.

   g) Megabytes capacity of a pen-drive and the number of photos you can store on it.

   h) People who lift an object and the strength they must do to load it.

2. Classify the relationship of proportionality (direct, inverse or non-proportional) between the following pairs of magnitudes:

   a) The number of kilos sold and the money raised.

   b) The number of workers who do work and time spent.

   c) A man's age and his height.

   d) The speed of a vehicle and the distance traveled in half an hour.

   e) The time it stays open a tap and the amount of water that is served.

   f) The flow of a tap and the time it takes to fill a reservoir.

   g) The number of pages of a book and its price.

3. Direct proportionality problems

How are proportionality problems solved? The method to solve proportionality problems depends on the proportionality relationship between the magnitudes comparing.

First of all, we are going to have to build a table with the data we have. So, we are going to have two fractions and we will have to build an equality with both fractions, *more or less*. We are going to see it with an example.

Example 2: If 3 kg of oranges cost € 2.64, what is the cost of 20 kg of them?

Solution: First of all, we are building a table with the data we have:

kg of oranges:	D	Euros:
3		2.64
20		X

We have also inserted a letter D between the two magnitudes we are comparing (kg and euros), meaning they are in direct proportion.

Now, we have two fractions, $\frac{3}{20}$ *and* $\frac{2.64}{x}$. As they are in **direct proportion**, we are building an equality with them, but writing at the left of the "=" sign, the fraction containing the "x":

$$\frac{2.64}{x} = \frac{3}{20}$$

As these two fractions are equivalent, we already know how to calculate "x", by "*Cross Multiplying*".

$$\frac{2.64}{x} = \frac{3}{20} \quad \Rightarrow \quad 3 \cdot x = 20 \cdot 2.64$$

$$3x = 52.8$$

$$x = \frac{52.8}{3} = 17.6 \ euros$$

Exercises	**3.** If 3 liters of petrol cost 3.45 €. How much will cost a) 5 liters? b) 23.5 liters?

96

4. If we travel 136 km in 1.5 hours driving at a constant speed.

a) How many km will we travel in 7.4 h?

b) How many hours do we need to travel 200 km?

5. Adrian finds that in each delivery of 500 bricks there are 20 broken bricks. How many bricks are broken in a delivery of 7500?

6. In a drink, 53 ml of fruit juice are mixed with 250 ml of water. How many liters of water are there in 30 l of that drink?

7. A car uses 25 liters of petrol to travel 176 miles. How far will the car travel using 44 liters of petrol?

8. Four kilos of potatoes cost € 1.80. How much do three kilos cost?

9. A machine produces 800 screws in 5 hours. How long will it take to produce 1000 screws?

10. A family drinks 2.5 L of milk daily. How many liters are consumed per week?
11. Complete in your notebook: If you need three baskets to take 15 melons, how many baskets will you need to take 195 melons?

12. A NGO delivers every day 48,000 kg of food using its 4 trucks. How many kg can they deliver one day if one of the trucks has broken down?

13. In a car factory, 380 cars have been built in 5 hours. How many cars will be produced in 12 hours, keeping the same ratio?

4. Inverse proportionality problems

The method to solve **inverse** proportionality problems has the same initial steps than direct proportionality method. It only has a little difference, before building the equality with the fractions.

We are going to see it with an example:

Example 3: Eight workers have to repair a street, and it takes them 90 minutes. Tomorrow, they must repair another similar street, but they are going to be 9 workers. How long will it take them?

Solution: First of all, we are building a table with the data we have:

Number of workers:	I	Number of minutes:
8		90
9		X

We have also inserted a letter I between the two magnitudes we are comparing (workers and minutes), meaning they are in inverse proportion.

Now, we have two fractions, $\dfrac{8}{9}$ *and* $\dfrac{90}{x}$. Now, we are building an equality with them, writing at the left of the *"="* sign, the fraction containing the *"x"*:   As they are in **inverse proportion**, we will *"UPSIDE-DOWN"* the other fraction (at the right of the *"="* sign).

$$\frac{90}{x} = \frac{9}{8}$$

As these two fractions are equivalent, we already know how to calculate *"x"*, by *"Cross Multiplying"*.

$$\frac{90}{x} = \frac{9}{8} \implies 9 \cdot x = 90 \cdot 8$$
$$9x = 720$$
$$x = \frac{720}{9} = 80 \ \text{min}$$

15. A truck that carries 3 tons need 15 trips to carry a certain amount of sand. How many trips are needed to carry the same amount of sand with another truck that carries 5 tons?

16. An automobile factory produces 8100 vehicles in 60 days. With the production rhythm unchanged. How many units will be made in one year?

17. A driver takes 3.5 hours to drive 329 km. How long will it take another trip in similar conditions as the previous one, but travelling 282 km instead?

18. Two hydraulic shovels make the trench for a telephone cable in ten days. How long will it take to make the trench with 5 shovels?

19. A farmer has grass to feed 20 cows for 60 days. If he had 30 cows, for how many days could he feed them?

20. At a fountain, it took 24 seconds to fill a 30-liter jar. How long does it take to fill a 50 liter container?

21. A pool has three similar taps. If opened two of them, pool gets full in 45 minutes. How long will it take filling it if you open all three?

22. Ten gardeners took eight days to repair all the trees in the streets of a town. How much would it have taken if they had been four gardeners only?

23. A gardener, working 8 hours per day, prepares his field in nine days. How long will it take him doing the same work, if he worked 12 hours a day?

24. Eight people collect all the oranges from an orchard in 9 hours. How long would it take them if they were 6 people?

5. Composite proportionality

We have already studied the proportionality relationship between two magnitudes. In some situations, there are three or more magnitudes having a proportionality relationship. These cases are named **composite proportionality**. Direct and inverse proportionality are both named *simple proportionality*.

How are composite proportionality problems solved?

The method to solve **composite** proportionality problems has lots of steps in common with simple proportionality one.

We are going to see it with an example.

Example 4: 50 cows consumed 4200 kg of grass in 7 days. How many kg of grass are needed to feed 20 cows for 15 days?

Solution: First of all, we are building a table with the data we have:

Cows:	Kg of grass:	Days:
50	4200	7
20	X	15

Now, we have to indicate the proportionality relationship of the magnitude containing the *"x"*, with each one of the others. We use two lines starting at the magnitude with the *"x"* and going to the others, writing on each line, the kind of proportionality between them:

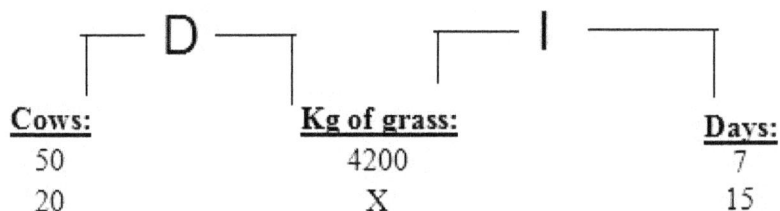

Cows:	Kg of grass:	Days:
50	4200	7
20	X	15

Now, we have three fractions, $\dfrac{50}{20}$, $\dfrac{4200}{x}$ *and* $\dfrac{7}{15}$. We are building an equality with them.

In composite proportionality problems, we write at the left of the "=" sign, the fraction containing the "x". Later, at the right of the "=" sign, we will write the product of the other two fractions, but writing them as in simple proportionality problems:

- If it is Direct proportional, we write the fraction.
- If it is Inverse proportional, we write *"UPSIDE-DOWN"* the fraction.

In our case:

$$\frac{4200}{x} = \frac{50}{20} \cdot \frac{15}{7}$$

Now, we can multiply the fractions at the right of the "=" sign. SIMPLIFY IF POSSIBLE!!!

$$\frac{50}{20} \cdot \frac{15}{7} = \frac{5}{2} \cdot \frac{15}{7} = \frac{\cdot 5 \cdot 3 \cdot 5}{2 \cdot 7} = \frac{75}{14}$$

And now: $\quad \dfrac{4200}{x} = \dfrac{75}{14}$

As these two fractions are equivalent, we already know how to calculate "x", by *"Cross Multiplying"*.

$$\frac{4200}{x} = \frac{75}{14} \quad \Rightarrow \quad 75 \cdot x = 4200 \cdot 14$$

$$75x = 58800$$

$$x = \frac{58800}{75} = 784 \ kg$$

Example 5: Five gardeners, working 12 hours a day, complete their gardening work in 10 days. How many gardeners would have been necessary to finish the same work, working 10 hours each day during 6 days?

Solution: First of all, we are building a table with the data we have:

Gardeners:	Hours per day:	Days:
5	12	10
X	10	6

Now, we have to indicate the proportionality relationship of the magnitude containing the *"x"*, with each one of the others. We use two lines starting at the magnitude with the *"x"* and going to the others, writing on each line, the kind of proportionality between them:

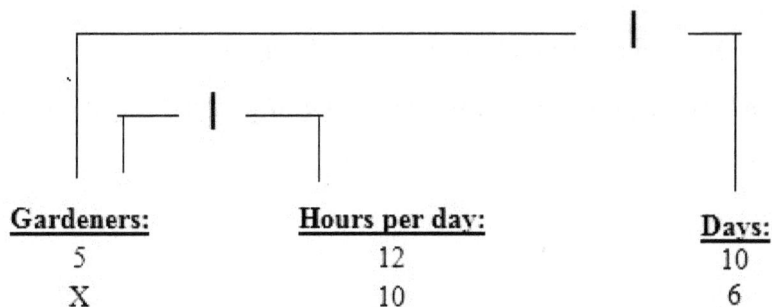

Gardeners:	Hours per day:	Days:
5	12	10
X	10	6

Now, we have three fractions, $\dfrac{5}{x}$, $\dfrac{12}{10}$ *and* $\dfrac{10}{6}$. We are building an equality with them.

In our case, both of the relationships are Inverse proportional, so, we write *"UPSIDE-DOWN"* both of them:

$$\frac{5}{x} = \frac{10}{12} \cdot \frac{6}{10}$$

Now, we can multiply the fractions at the right of the *"="* sign.   SIMPLIFY IF POSSIBLE!!!

$$\frac{10}{12} \cdot \frac{6}{10} = \frac{6}{12} = \frac{2 \cdot 3}{2 \cdot 2 \cdot 3} = \frac{1}{2}$$

And now: $\dfrac{5}{x} = \dfrac{1}{2}$

As these two fractions are equivalent, we already know how to calculate *"x"*, by *"Cross Multiplying"*.

$\dfrac{5}{x} = \dfrac{1}{2}$ $\Rightarrow$ $1 \cdot x = 5 \cdot 2$

$$x = 10 \, people$$

25. In a manufacturing workshop, with 6 sewing machines, 600 jackets were manufactured in 10 days. How many jackets would be manufactured with 5 machines in 15 days?

26. An industrial washer, working 8 hours a day for five days, has washed 1000 kg of clothes. How many kg of clothes will be washed in 12 days working 10 hours a day?

27. Five interviewers, working 8 hours a day, complete data for a market survey in 27 days. How long will it take them doing the same work if they were 9 people working 10 hours each day?

28. The garden of a park has been cleaned by 3 gardeners working for 15 days and working 8 hours daily. In another park, they want to create another similar garden having the same extension. As they want to finish in 4 days, they have called nine gardeners. How many hours must work daily?

6. Percentages

A *percentage* is a ratio of a number to 100. A percentage is expressed using the symbol %.

A percentage is also equivalent to a fraction with a denominator of 100.

For example, 65% is equivalent to the fraction $\dfrac{65}{100}$.

6.1. Percentage as a comparison of two numbers

To find the percentage as a comparison of two numbers

1. Write the ratio of the first number to the base number.
2. Multiply by 100 this fraction and add the symbol %.

Example 6: What percentage of the region is shaded?

a)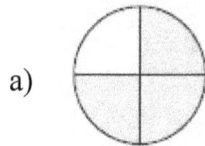

b)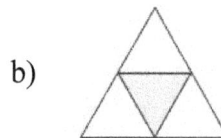

Solution: a) $\dfrac{3}{4} \cdot 100 = \dfrac{3 \cdot 100}{4} = \dfrac{300}{4} = \boxed{75\%}$.

b) $\dfrac{1}{4} \cdot 100 = \dfrac{1 \cdot 100}{4} = \dfrac{100}{4} = \boxed{25\%}$.

29. What percentage of the region is shaded?

a)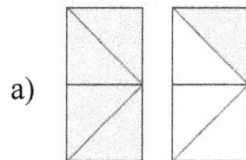

b)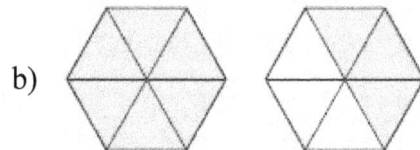

30. Write an exact percentage for these comparisons. Some of them might be calculated directly:

a) 62 of 100   b) 52 per 100   c) 32 to 100   d) 37 to 100   e) 28 per 50

f) 17 per 50   g) 12 of 25   h) 21 to 25   i) 11 per 20   j) 13 per 20

k) 13 to 10   l) 450 to 120   m) 313 of 313   n) 92 to 92   ñ) 30 to 12

o) 44 to 16   p) 85 to 200   q) 65 to 200   r) 15 per 40   s) 83 per 500

t) 70 per 80   u) 98 per 80.

31. It is estimated that 2% of the U.S. population has red hair. This indicates that ___ out of 100 people are redheads.

32. In a recent election there was a 73% turnout of registered voters. This indicates that ___ out of 100 registered voters turned out to vote.

33. In a recent mail-in election, 18 out of every 100 eligible voters did not participate. What percentage of the eligible voters exercised their right to vote?

34. Of the people who use mouthwash daily, 63 out of 100 report fewer cavities. Of every 100 people who report, what percentage do not report fewer cavities?

35. At a football game, 22 children are among the first 100 fans to enter. What percentage of the first 100 fans are children?

36. At the last soccer match of the season, of the first 100 tickets sold, 77 were student tickets. What percentage were student tickets?

37. Write the ratio of 8 to 5 as a percent.

38. During a campaign to lose weight, the 180 participants lost a total of 4158 lb. If they weighed collectively 37,800 lb before the campaign, what percentage of their weight was lost?

39. James has $500 in his savings account. Of that amount, $35 is interest that was paid to him. What percentage of the total amount is the interest?

40. According to the U. S. Census Bureau, in 2007 one out of every three women aged 25 to 29 had a bachelor's degree or higher. What percentage of women 25 to 29 had a bachelor's degree?

41. According to the U. S. Census Bureau, in 2007, 13 out of every 50 men aged 25 to 29 had a bachelor's degree or higher. What percentage of men 25 to 29 had a bachelor's degree?

42. Mickie bought a TV and makes monthly payments on it. Last year, she paid a total of $900. Of the total that she paid, $180 was interest. What percentage of the total was interest?

7. Solving problems with percentages

To solve problems involving percentages, we will use the following expression:

$$Portion = \frac{Percentage}{100} \cdot Whole$$

Anyway, problems with percentages can also be solved as a direct proportionality. We are seeing it in the following examples.

7.1. Calculating the portion

Example 7: What is the 35% of 300?

Solution: We will solve it by applying the percent expression and as a direct proportionality.

Method 1: Percentage expression:	Method 2: As a direct proportionality:
- Portion = x? Calculate: - Percentage = 35 - Whole = 300 $x = \dfrac{35 \cdot 300}{100} = \boxed{105}$. $x = \dfrac{35}{100} \cdot 300$	**Whole:** $\boxed{\text{D}}$ **Portion:** 100 35 300 x $\dfrac{100}{300} = \dfrac{35}{x} \rightarrow x \cdot 100 = 300 \cdot 35; \; x = \dfrac{35 \cdot 300}{100} = \boxed{105}$.

<table>
<tr><td rowspan="4">Exercises</td><td colspan="4">43. Calculate:</td></tr>
<tr><td>a) 60% of 80</td><td>b) 30% of 91</td><td>c) 55% of 72</td><td>d) 130% of 90</td></tr>
<tr><td>e) 140% of 70</td><td>f) 17.5% of 70</td><td>g) 57.5% of 110</td><td>h) 45.5% of 80</td></tr>
<tr><td>i) 17.2% of 55</td><td>j) 6.14% of 350</td><td>k) 12.85% of 980</td><td>l) 16 % of 3522</td></tr>
</table>

7.2. Calculating the whole

Example 8: 54% of what number is 108?

Solution: We will solve it by applying the percent expression and as a direct proportionality.

Method 1: Percentage expression:	Method 2: As a direct proportionality:
- Portion = 108 Isolate x: - Percentage = 54 - Whole = x? $108 \cdot \dfrac{100}{54} = x = \dfrac{10800}{54} =$ $108 = \dfrac{54}{100} \cdot x$ $\qquad = \boxed{200}$.	**Whole:** $\boxed{\text{D}}$ **Portion:** 100 54 x 108 $\dfrac{100}{x} = \dfrac{504}{108} = \dfrac{1}{2} \rightarrow x \cdot 1 = 2 \cdot 100 = \boxed{200}$.

44. Calculate and complete:

    a) 80% of _____ is 32.   b) 39% of _____ is 39.   c) 497.8 is 76% of ___ .

    d) 162 is 18% of ___.   e) 39% of ___ is 105.3.   f) 73% of ___ is 83.22.

    g) 124% of ___ is 328.6.   h) 205% of ___ is 750.3.

All together. Calculating percentages, portions and wholes

45. What percentage of 85 is 41? Round to the nearest tenth of a percent.

46. What percentage of 666 is 247? Round to the nearest tenth of a percent

47. Eighty-two is 24.8% of what number? Round to the nearest hundredth.

48. Forty-one is 35.2% of what number? Round to the nearest hundredth.

49. Thirty-two and seven tenths percent of 695 is what number?

50. Seventy-three and twelve hundredths percent of 35 is what number?

51. Thirty-seven is what percentage of 156? Round to the nearest tenth of a percent.

52. Two hundred thirty-two is what percentage of 124? Round to the nearest tenth of a percent.

53. The cost of a certain model of Ford is 120% of what it was 5 years ago. If the cost of the automobile 5 years ago was $20,400, what is the cost today?

54. In a 1st ESO group, 10 students played football, 15 basketball and 5 tennis. What is the percentage of the students playing football?

55. In a study of 615 people, 185 said they jog for exercise. What percentage of those surveyed jog? Round to the nearest whole percent.

56. Based on a survey, approximately 77% of TV sets in the United States receive cable. If there are about 304,000,000 sets in the United States, how many do not get cable? Round to the nearest million.

57. The town of Verboort has a population of 17,850, of which 48% is male. Of the men, 32% are 40 years or older. How many men are there in Verboort who are younger than 40?

Exercises

7.3. Percentage changes

As you have seen, percent changes can be calculated by using previous expression. Anyway, there is an easier expression you can use. It is the following:

Expression to calculate percentage changes:

$$F.A. = I.A. \cdot \left(1 \pm \frac{\%}{100}\right)$$

Where: - F.A.: Final amount.

       - I.A.: Initial amount.

       - %: Percentage change.

       - $\pm$: Use (+) when you consider an increase and a (-) in decreases.

Example 9: Prices of houses have increased 20% from five years ago. In that moment, I paid 120,000 euros for my house. How much would I have to pay today?

Solution:

    % = 20;   I.A. = 120,000; F.A. = x?; plus or minus? As prices have increased, we use <u>plus</u> $\rightarrow$

$$F.A. = 120000 \cdot \left(1 + \frac{20}{100}\right) = 120000 \cdot (1 + 0.2) = 120000 \cdot 1.2 = \boxed{144,000 \text{ euros}}.$$

Example 10: In the shop where I am used to buying my clothes, there are 15% sales today. I have paid 25.5 euros for T-shirt. What was its price before sales?

Solution:

    % = 15; F.A. = 25.5; I.A. = x?; plus or minus? As prices have decreased, we use <u>minus</u> $\rightarrow$

$$F.A. = I.A. \cdot \left(1 \pm \frac{\%}{100}\right) \rightarrow 25.5 = I.A. \cdot \left(1 - \frac{15}{100}\right); \quad 25.5 = I.A. \cdot 0.85; \quad \frac{25.5}{0.85} = I.A. = \boxed{30 \ euros}$$

Example 11: When John began to work in his company, his salary was 1100 euros per month. From that moment, his salary has increased twice, first, 15%, and later, 20%. But, this year, as there are problems with Spanish economy, he has had his salary decreased 35%. What is his salary today?

Solution: We might think *"it increased 15 + 20 = 35%, and it decreased 35%, so, he should have the same initial salary"*.

We are going to see this is not true. As there are more than one percent change, we will use more than one parentheses in above expression:

$$F.A. = 1100 \cdot \left(1 + \frac{20}{100}\right) \cdot \left(1 + \frac{15}{100}\right) \cdot \left(1 - \frac{35}{100}\right) = 1100 \cdot 1.2 \cdot 1.15 \cdot 0.65 = \boxed{986.7 \text{ euros}}.$$

You can see his salary has decreased. Be careful in the future, don't forget this!!!

Exercises

58. I am taking part in a program to lose weight. I have already lost a 20%. If my weight now is 60 kg, what was my weight before the program?

59. Today there are 20% sales in a shop near my house. This was a T-shirt's label: "Before 25 €; Today 22.5 €". Do you think this label is correct? Why?

60. During ten last years, prices of houses, first increased 10%, later increased 20% and, this year, they have decreased 30%. Ten years ago, I paid 150,000 € for my house. Check if its price has changed from them. Has it increased or decreased?

Review exercises

1. Classify the relationship of proportionality (direct, inverse or non-proportional) between the following pairs of magnitudes:

    a) Speed and time at a constant speed motion.

    b) Space and time in motion with constant velocity.

    c) Number of persons sharing a cake and size of portion corresponding to each one.

    d) Number of hours a student watches television and hours of study.

    e) Amount of money that you save and your weight.

    f) Number of correct answers and number of failures in an exam.

    g) Number of workers and time it takes to build a wall.

    h) Number of people and amount of food eaten.

    i) Number of people involved in buying a gift and money they bring.

    j) Number of laborers and time taken to finish their work.

2. Complete the table of the cost in € and the liters of petrol bought.

Petrol in litres	1		5	7	
Cost in €		40	6,2		50

3. A phone call costs € 0.25 each 2 minutes or fraction rounded to the seconds. Complete the table.

Call length in minutes	7,5		13 minutes 25 seconds
Cost in €		12	

4. Richard earns £17.5 for working 7 hours. How much will he earn for working 9 hours?

5. Two pumps take 5 days to empty a pool. How long will 5 pumps take to empty the same pool?

6. We have paid for 7 nights in "Hotel Los Llanos" 364 €. How much will we pay for 3 nights? How much for 15 nights?

7. It takes 12 hours for 3 bricklayers to build a wall. How long will it take for 5 bricklayers?

8. A 25 kg tin of paint covers 70 m^2 of wall. How many kg would be needed to cover 53 m^2 of wall?

9. A company need 33 workers to pack a its production in 25 days. If the total production needs to be packed in 15 days, how many <u>extra</u> workers do they need?

10. If I ride my bicycle at an average speed of 15 km/h, I travel a distance of 22 km in a certain period of time. If the speed is 17 km/h, how far will I travel?

11. A tap opened 9 hours during 8 days has thrown 5400 liters of water. How many liters will throw during 18 days, 8 hours per day?

12. If 18 lorries carry 1200 containers in 12 days, How long will 24 lorries need to carry 1600 containers?

13. We have used 2 cans with 1 Kg of paint, to paint a wall with dimensions 8 m x 2.5 m. How many cans, with 5 kg of paint, will we need to paint another wall with dimensions 50 m x 2 m?

Percentages

14. The population of a town is 652 000 and 35% of it live in the central district. How many of the people live in this district?

15. During a sale the price of a television set is 150 € which is 75 % of the usual price, what is the original price?

16. 6% of the population in Murcia are immigrants and there are 9900 immigrants living in the city, what is the population in Murcia?

17. Some clothes are priced at 68 € and there is a discount of 7%, what is the final price?

18. I have bought a pair of jeans for 33 €, the VAT is 10 %, what was the price before?

19. The price of an electric oven before taxes is 560 € plus 17% VAT and the salesman offers a 12% discount, what is the final price?

20. Last year there were 1560 employees in a company, this year 312 new people have been employed. What has been the percentage of increase in the staff of the company?

21. I have paid 161 € for a coat and the original price was 230 €. What is the % discount?

22. In shop A, a mobile phone costs 99 € plus VAT (16%). In shop B, the same model costs 114.90 € (VAT included). In what shop is the mobile phone cheaper?

23. At a sporting event that brings together 750 athletes, 30% of them are American, 8% Asian, 16% African, and the remainder European. How many European athletes take part in the match?

24. What's the total number of guests who attend a wedding, knowing that there are 33 men and 45% are women?

25. The price of a book after an increase of 20 % is 4.20 €. How much did it cost before the rise?

Unit 6.- Algebraic expressions

1. Algebraic expressions

A *variable* is a symbol that represents a number. We usually use letters such as x, n, p, t for variables.

Letters are useful if we want to operate with an unknown number instead with a particular one. Let us look at some examples:

- We say that *s* represents the side of a square, then *s* represents a number and:

    4s Is the perimeter of the square $s \times s = s^2$ Is the area of the square

When letters express numbers they have the same operating properties. The part of mathematics that deals with the study of the expressions with letters and numbers is called **algebra**.

An *algebraic expression* is a combination of numbers, variables (algebraic quantities) and arithmetic operators: +, −, x, ÷.

For example, these are expressions: 3x   x+3   2(x − 5)   x   3x − 2

If Mark weighs 80 kg and he gains n kg, the new weight is 80 + n

Look at these examples:

The triple of a number:	3n
The triple of a number minus five units:	3n-5
The following number:	x+1
The preceding number:	x-1
An even number:	2x
An odd number:	2x+1

Exercises

1. Find the expressions:

a) I start with x, double it and the subtract 10.

b) I start with x, add 3 and then square the result.

c) I start with x, take away 5, double the result and then divide by 5.

d) I start with x, multiply by 5 and then subtract 3.

e) I start with x, add y and then double the result.

f) I start with n, square it and then subtract n.

g) I start with x, add 2 and then square the result.

h) A brick weighs x kg. How much do 6 bricks weigh? How much do n bricks weigh?

2. Write using algebraic language:

a) The double of a number plus seven.

b) The triple of a number minus one.

c) The double of the sum of a number and three.

d) The half of the difference between a number and four.

3. Calling '*a*' the age of a person today, write an expression for:

a) The age he/she will be in 5 years.

b) The age he/she was 7 years ago

c) The age he/she will be after living the same time again.

4. Read and complete:

- Pablo´s salary is x euros.

- The gerent of the firm earns a double quantity than Pablo.

- The engineer earns 400 € less than the gerent.

- The person who cleans the office earns 10% less than Pablo.

Pablo	Gerent	Engineer	Cleaning service

2. Numerical value of an expression

Numerical value of an expression, for a given value of the variable, is the result of substituting the variable by the given value. If *P(x)* is an expression, its numerical value when $x = a$, is denoted by *P(a)*.

Example 1: Evaluate $5x^2 + 3$ when $x = 2$.

Solution: As $P(x) = 5x^2 + 3$, $P(2) = 5 \cdot 2^2 + 3 = 5 \cdot 4 + 3 = 20 + 3 = \boxed{23}$.

5. Calculate the numerical value of the algebraic expression $2x^2 - 1$ when:

   a) $x = 3$ b) $x = 1$ c) $x = (-2)$ d) $x = (-5)$

6. Calculate the numerical value of the expression $2x - y^2 + 3z + 1$ when:

   a) $x = 1$, $y = 0$, $z = (-1)$ b) $x = 2$, $y = 1$, $z = 0$ c) $x = 3$, $y = (-2)$, $z = 1$

7. Work out the value of the algebraic expression $2a - 3b$ when $a = 5$ and $b = (-1)$

8. Evaluate the algebraic expression $2x^2 - 5$ when:

   a) $x = 2$ b) $x = 0$ c) $x = (-1)$ d) $x = 1/3$.

9. Calculate the numerical value of the algebraic expression $2a + 3b^2 - 5$ when:

   a) $a = 2$, $b = (-1)$ b) $a = (-1)$, $b = 2$.

3. Monomials and operations

A monomial consists of the product of a known number **(coefficient)** by one or several letters with exponents that must be constant and positive whole numbers **(literal part)**.

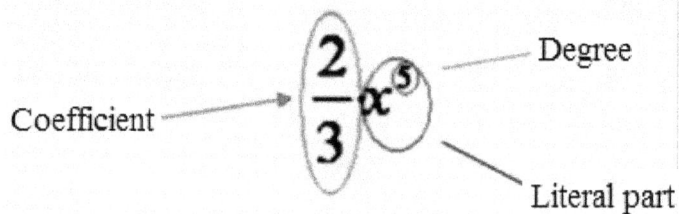

Two or more monomials are said to be **like monomials** if they have the same literal part.

Examples: The expressions $21x^2$, $8x^3$, $\frac{2}{5}x^2$, $-3x^5$ are monomials, and $21x^2$ and $\frac{2}{5}x^2$ are *like monomials*.

12. Write a monomial with variable x that verifies the following conditions:

a) Degree 2 and coefficient (-4).

b) Degree 0 and coefficient 1.

c) A monomial being like to this $4x^3$ and with coefficient 3.

d) Degree 3 and coefficient 1/3.

3.1. Addition and subtraction of monomials

Two monomials can only be added or subtracted if they are **like monomials**. So, they must have the same literal part. In this case, we add or subtract the coefficients and we **leave the literal part unchanged**. When the literal part is different, the addition is left indicated.

The result of the **addition** or **subtraction** of two **like monomials** is another like monomial whose coefficient is the sum or subtraction of the coefficients.

$$ax^n \pm bx^n = (a \pm b)x^n.$$

13. Calculate:

a) $3x^2 + 4x^2 - 5x^2$

b) $6x^3 - 2x^3 + 3x^3$

c) $x^5 + 4x^5 - 7x^5$

d) $-2x^4 + 6x^4 + 3x^4 - 5x^4$

e) $7x + 9x - 8x + x$

f) $2y^2 + 5y^2 - 3y^2$

g) $3x^2y - 6x^2y + 5x^2y$

h) $4xy^2 - xy^2 - 7xy^2$

i) $2a^6 - 3a^6 - 2a^6 + a^6$

j) $ab^3 + 3ab^3 - 5ab^3 + 6ab^3 - 4ab^3$

k) $7xy^2z - 2xy^2z + xy^2z - 6xy^2z$

l) $-x^3 + 5x - 2x + 3x^3 + x + 2x^3$

m) $x^4 + x^2 - 3x^2 + 2x^4 - 5x^4 + 8x^2$

n) $3a^2b - 5ab^2 + a^2b + ab^2$

o) $\frac{7}{3}x^2 + \frac{4}{3}x^2$

p) $12x^5 - x^5 - 4x^5 - 2x^5 - 3x^5$

117

q) $\dfrac{7}{4}x^5 - \dfrac{1}{4}x^5$

r) $x^2y^2 - 5x^2y^2 - (3x^2y^2 - 4x^2y^2) - 8x^2y^2$

s) $x^2 + \dfrac{x^2}{3}$

t) $x^2 + x^2$

u) $\dfrac{1}{2}x^3 - \dfrac{5}{2}x^3 + \dfrac{3}{2}x^3$

v) $-(ab^3 + a^3b) - 3a^3b + 5ab^3 - (a^3b - 2ab^3)$

w) $7x^3 - \dfrac{1}{2}x^3 - \dfrac{5}{2}x^3 + 2x^3 + \dfrac{3}{2}x^3$

x) $2a^2b + 5a^2b - \dfrac{2}{3}a^2b - a^2b + \dfrac{a^2b}{2}$

14. Calculate:

a) $x^4 + x^3 - x^2 + x^4 - x^3 - x^2$

b) $3x^3 - (5x^3 + 7x^3)$

c) $9x^2 - (x^2 - 3x^2)$

d) $9xy^2 - 5xy^2 + 7xy^2 - xy^2$

15. Collect like terms to simplify each expressions:

a) $x^2 + 3x - x + 3x^2$

b) $3x - 4 - (x + 1)$

c) $5y + 3x + 2y + 4x$

d) $(2x + 3) - (5x - 7) - (x + 1)$

e) $\dfrac{2}{3}x^2 + \dfrac{1}{2}x - \dfrac{3}{2}x^2 - \dfrac{1}{5}x + 2$

3.2. Product and quotient of two monomials

To **multiply** two or more monomials, we must multiply, <u>separately</u>, their coefficients and literal parts.

Remember the method to multiply powers having the same base:

$$x^m \cdot x^n = x^{(m+n)}.$$

Example 2: Multiply: a) $5x^2 \cdot 7x^3$ b) $\dfrac{3}{2}x^2 \cdot 4x^3$

<u>Solution:</u>

a) $5x^2 \cdot 7x^3 = \begin{cases} 5 \cdot 7 = 35 \\ x^2 \cdot x^3 = x^{(2+3)} = x^5 \end{cases} = \boxed{35x^5}$

b) $\dfrac{3}{2}x^2 \cdot 4x^3 = \begin{cases} \dfrac{3}{2} \cdot 4 = \dfrac{3}{2} \cdot \dfrac{4}{1} = \dfrac{12}{2} = 6 \\ x^2 \cdot x^3 = x^{(2+3)} = x^5 \end{cases} = \boxed{6x^5}$

118

To **divide** two or more monomials, we must divide, separately, their coefficients and literal parts.

Remember the method to divide powers having the same base:

$$x^m : x^n = x^{(m-n)}.$$

Example 3: Divide: a) $15x^6 : 3x^2$ b) $\dfrac{3}{2}x^7 : \dfrac{9}{8}x^3$

Solution:

a) $15x^6 : 3x^2 = \begin{cases} 15 : 3 = 5 \\ x^6 : x^2 = x^{(6-2)} = x^4 \end{cases} = \boxed{5x^4}$

b) $\dfrac{3}{2}x^7 : \dfrac{9}{8}x^3 = \begin{cases} \dfrac{3}{2} : \dfrac{9}{8} = \dfrac{3}{2}\cdot\dfrac{8}{9} = \dfrac{3\cdot2\cdot2}{2\cdot3\cdot3} = \dfrac{2\cdot2}{2} = \dfrac{4}{3} \\ x^7 : x^3 = x^{(7-3)} = x^4 \end{cases} = \boxed{\dfrac{4}{3}x^4}$

Exercises

16. Calculate:

a) $x^2 \cdot x^3 \cdot x^6$ b) $3x^2 \cdot 2x^4 \cdot 5x$ c) $-2x^2 \cdot 5x^2 \cdot x^2$ d) $\dfrac{2}{5}x^3 \cdot 4x \cdot (-3x^2)$

e) $(12x^7):(4x^2)$ f) $(-10x^5 \cdot y^3):(5x^4\,y)$ g) $\dfrac{18x^6}{6x^2}$ h) $\dfrac{20x^4y^3}{5xy}$

17. Calculate:

a) $3x^2 \cdot 4x^3$ b) $2x^3 \cdot 4x^3 \cdot 3x^3$

c) $x^3 \cdot x^3$ d) $-2x^4 \cdot 3x^3$

e) $7x \cdot (-8x^2)$ f) $(-3y^2) \cdot (-2y^3)$

g) $3x^2y \cdot 6xy^3$ h) $\dfrac{3}{4}x^2 \cdot \dfrac{5}{2}x^3$

i) $4a^3b^2 \cdot a^2b \cdot 7ab$ j) $-\dfrac{1}{2}a^3 \cdot \dfrac{5}{3}a^4$

k) $2a^6 \cdot 3a^6 \cdot 2a^6$ l) $\dfrac{2}{5}x^3 \cdot \left(-\dfrac{3}{2}x\right)$

m) $ab^3 \cdot (-3a^2b) \cdot 5a^3b$

n) $x^2 \cdot \dfrac{1}{3}x^5$

o) $-ab^2c^3 \cdot (-3a^2bc) \cdot 3abc$

p) $(6x^4) : (2x^2)$

q) $\dfrac{12a^6}{3a^3}$

r) $15x^4 : (-3x)$

s) $\dfrac{-14x^7}{7x^2}$

t) $-8x^4 : (-4x^3)$

u) $\dfrac{5x^7y^3}{x^2y}$

v) $(-18x^4) : (6x^3)$

w) $\dfrac{-12a^5b^4c^6}{2a^3b^2c}$

x) $2x^4 \cdot 6x^3 : 4x^2$

y) $\dfrac{3a^5b \cdot (-12a^4b^2)}{4a^3b^2}$

z) $27x^4 : (-9x^3) \cdot (-2x^2)$

18. Calculate:

a) $15x \cdot 2x^2 : x + 20x \cdot x^3 : 4x^2$

b) $2x^2 \cdot 2 \cdot 3x^3 - 5x^4 \cdot 4x^2 : 2x$

c) $(-16xy^2 + 2xy^2) : (3xy + 4xy)$

d) $(3x + 8x - 7x) \cdot (2x^2 + 5x^2 - 6x^2)$

4. Polynomials and their operations

When we want to add or subtract unlike monomials, we cannot simplify them and we must leave these operations indicated. These sums or subtractions of monomials are named polynomials.

For example, we have the polynomial $9x^3 + 8x^2 + 6x + 12$.

A *polynomial* in the unknown x is an algebraic expression given by the sum or subtraction of two or more monomials in the same unknown.

More definitions:

> · We name *degree* of a polynomial to the highest of the degrees of the monomials in it.
>
> · A polynomial is *complete* when it has all the terms, from the term with degree 0 to the term with the highest degree.
>
> · A polynomial is *ordinated* when the degrees of the terms are increasing or decreasing. Usually, monomials are ordinated in a decreasing order.

Polynomials are named as $A(x)$, $B(x)$, etc. indicating in the parenthesis the unknown. So, the following expressions are polynomials.

Polynomial	Number of terms	Independent term	Degree	Complete	Ordinated
$A(x) = 5x^4 - 3x^2 + x - 1$	4	-1	4	Not	Yes
$B(x) = 2 - 5x^2 + 4x$	3 (trinomial)	2	2	Yes	Not
$C(x)\ 3x - 7$	2 (binomial)	-7	1	Yes	Yes

4.1. Sum and subtraction of polynomials

Adding or subtracting polynomials consists simply in adding or subtracting their like monomials. Given two polynomials $A(x) = x^3 - 2x^2 - 7x - 4$ and $B(x) = -x^4 + 8x^2 + 7x + 2$, we can calculate $A(x) + B(x)$ by writing one below the other, so that we collocate in the same column the like monomials. After that, we add the like monomials.

$$
\begin{array}{rl}
A(x) = & x^3 - 2x^2 - 7x - 4 \\
B(x) = -x^4 & + 8x^2 + 7x + 2 \\
\hline
A(x) + B(x) = -x^4 + x^3 + 6x^2 & - 2
\end{array}
$$

Opposite polynomial of a given polynomial is another one that has the same monomials but with opposite coefficients (opposite signs).

If B(x) = - x^4 + $8x^2$ + 7x + 2, its opposite polynomial is -B(x) = x4 - 8x2 - 7x - 2.

Do you agree that $3 - 2 = 3 + (-2)$? So, in the same way, we can calculate A(x) - B(x) = A(x) + [-B(x)]:

$$
\begin{aligned}
A(x) = \quad & x^3 - 2x^2 - 7x - 4 \\
-B(x) = x^4 \quad & \quad - 8x^2 - 7x - 2 \\
\hline
A(x) - B(x) = x^4 + & x^3 - 10x^2 - 14x - 6
\end{aligned}
$$

Another method to add or subtract polynomials is the application of the sign's rule for addition or subtraction of like monomials.

First of all, we remove the parentheses:

- If it is preceded by a + sign, we conserve the original signs.

- If it is preceded by a − sign, we change the sign inside the parenthesis.

We are using this method to calculate again A(x) + B(x) and A(x) - B(x):

A(x) + B(x) = (x^3 - $2x^2$ - 7x - 4) + (- x^4 + $8x^2$ + 7x + 2) = x^3 - $2x^2$ - 7x - 4 - x^4 + $8x^2$ + 7x + 2 = $\boxed{-}$ $\boxed{x^4 + x^3 + 6x^2 - 2}$

A(x) - B(x) = (x^3 - $2x^2$ - 7x - 4) - (- x^4 + $8x^2$ + 7x + 2) = x^3 - $2x^2$ - 7x - 4 + x^4 - $8x^2$ - 7x-2 = $\boxed{x^4 + x^3 -}$ $\boxed{10x^2 - 14 - 6}$

19. Given the following polynomials:

$P(x) = 2x^3 - 3x^2 + 4x - 2$

$Q(x) = x^4 - x^3 + 3x^2 + 4$

$R(x) = 3x^2 - 5x + 5$

$S(x) = 3x - 2$

Calculate:

a) $P(x) + Q(x)$ (Sol: $x^4 + x^3 + 4x + 2$)

b) $P(x) + R(x)$ (Sol: $2x^3 - x + 3$)

c) $P(x) - R(x)$ (Sol: $2x^3 - 6x^2 + 9x - 7$)

d) $Q(x) - S(x)$ (Sol: $x^4 - x^3 + 3x^2 - 3x + 6$)

e) $P(x) + S(x)$ (Sol: $2x^3 - 3x^2 + 7x - 4$)

f) $Q(x) + R(x)$ (Sol: $x^4 - x^3 + 6x^2 - 5x + 9$)

g) $Q(x) + S(x)$ (Sol: $x^4 - x^3 + 3x^2 + 3x + 2$)

h) $P(x) - S(x)$ (Sol: $2x^3 - 3x^2 + x$)

i) $S(x) - P(x)$ (Sol: $-2x^3 + 3x^2 - x$)

j) $P(x) - Q(x) + R(x)$ (Sol: $-x^4 + 3x^3 - 3x^2 - x - 1$)

k) $Q(x) - [R(x) + S(x)]$ (Sol: $x^4 - x^3 + 2x + 1$)

l) $S(x) - [R(x) - Q(x)]$ (Sol: $x^4 - x^3 + 11x - 3$)

4.2. Expansion of brackets.

Multiplying a number by an addition is equal to multiplying by each adding number and then to adding the partial products.

Example 4: Expand the following expressions in brackets:

a) $2(x + 6)$ b) $3(2x - 7)$ c) $3x(3x+2)$

Solution:

a) $2(x + 6) = 2x + 12$

b) $3(2x - 7) = 6x - 21$

c) $3x(3x+2) = 9x^2 + 6x$

20. Copy and fill in the missing terms:

a) $5(3x - 4) = \boxed{...} - 20$ b) $x(x - 2) = \boxed{...} - 2x$

c) $a(4 - a) = \boxed{...} - a^2$ d) $7x(3x - y) = \boxed{...} - 7xy$

21. Expand:

a) $x \cdot (x +3)$ b) $x \cdot (3x - 5)$ c) $x^2 \cdot (4x^2 + 7x - 2)$ d) $7 \cdot (5x^2 - 3)$

e) $(-3x) \cdot (2x + 5)$ f) $x \cdot (7x - 1)$ g) $2 \cdot (x^2 - 2x + 1)$ h) $(-2x) \cdot (5x + 1 - 3x^2)$

22. Calculate:

a) $(3x^4 - 2x^3 + 2x^2 + 5) \cdot 2x^2$ (Sol.: $6x^6 - 4x^5 + 4x^4 + 10x^2$)

b) $(-2x^5 + 3x^3 - 2x^2 - 7x + 1) \cdot (-3x^3)$ (Sol.: $6x^8 - 9x^6 + 6x^5 + 21x^4 - 3x^3$)

c) $\left(\dfrac{2}{3}x^3 - \dfrac{3}{2}x^2 + \dfrac{4}{5}x - \dfrac{5}{4} \right) \cdot 12x^2$ (Sol.: $8x^5 - 18x^4 + \dfrac{48}{5}x^3 - 15x^2$)

d) $\left(\dfrac{1}{2}ab^3 - a^2 + \dfrac{4}{3}a^2b + 2ab \right) \cdot 6a^2b$ (Sol.: $3a^3b^4 - 6a^4b + 8a^4b^2 + 12a^3b^2$)

23. Expand:

a) $5 \cdot (1 + x)$ b) $(-4) \cdot (2 - 3a)$ c) $3a \cdot (1 + 2a)$

d) $x^2 \cdot (2x - 3)$ e) $x^2 \cdot (x + x^2)$ f) $2a \cdot (a^2 - a)$

Sol.: a) $5 + 5x$; b) $-8 + 12a$; c) $3a + 6a^2$; d) $2x^3 - 3x^2$; e) $x^3 + x^4$; f) $2a^3 - 2a^2$

24. Expand and simplify as much as possible:

a) x + 2(x + 3)   b) 7x – 3(2x – 1)   c) 4 · (a + 2) – 8

d) 3 · (2a – 1) – 5a   e) 2(x + 1) + 3(x – 1)   f) 5 · (2x – 3) – 4 · (x – 4)

Sol.: a) 3x + 6; b) x + 3; c) 4a; d) a – 3; e) 5x – 1; f) 6x + 1.

Exercises

4.3. Product of two polynomials

To **multiply** two polynomials, we must multiply each monomial of the first one by all the monomials of the second one, or vice versa.

For example, given the polynomials $A(x) = x^3 + 2x^2 - x + 2$ y $B(x) = 4x - 2$, we calculate $A(x) \times B(x)$:

$$
\begin{array}{r}
A(x) = \quad x^3 + 2x^2 - x + 2 \\
\times \quad B(x) = \quad 4x - 2 \\
\hline
-2x^3 - 4x^2 + 2x - 4 \\
4x^4 + 8x^3 - 4x^2 + 8x \\
\hline
A(x) \cdot B(x) = 4x^4 + 6x^3 - 8x^2 + 10x - 4
\end{array}
$$

As we have already said, we can multiply two polynomials by applying the distributive property. We mean:

$$A(x) \times B(x) = (x^3 + 2x^2 - x + 2) \times (4x - 2) = x^3(4x - 2) + 2x^2(4x - 2) - x(4x - 2) + 2(4x - 2) =$$

$$= 4x^4 - 2x^3 + 8x^3 - 4x^2 - 4x^2 + 2x + 8x - 4 = \boxed{4x^4 + 6x^3 - 8x^2 + 10x - 4}$$

25. Given the following polynomials:

$$P(x) = 2x^3 - 3x^2 + 4x - 2$$

$$Q(x) = x^4 - x^3 + 3x^2 + 4$$

$$R(x) = 3x^2 - 5x + 5$$

$$S(x) = 3x - 2$$

Calculate the following products:

a) $P(x) \cdot R(x)$ (Sol: $6x^5 - 19x^4 + 37x^3 - 41x^2 + 30x - 10$)

b) $P(x) \cdot S(x)$ (Sol: $6x^4 - 13x^3 + 18x^2 - 14x + 4$)

c) $S(x) \cdot P(x)$ (Sol: Ídem)

d) $P(x) \cdot P(x)$ (Sol: $4x^6 - 12x^5 + 25x^4 - 32x^3 + 4x^2 - 8x + 4$)

e) $Q(x) \cdot S(x)$ (Sol: $3x^5 - 5x^4 + 11x^3 - 6x^2 + 12x - 8$)

f) $[Q(x)]^2$ (Sol: $x^8 - 2x^7 + 7x^6 - 6x^5 + 17x^4 - 8x^3 + 24x^2 + 16$)

g) $R(x) \cdot S(x)$ (Sol: $9x^3 - 21x^2 + 25x - 10$)

h) $[R(x)]^2$ (Sol: $9x^4 - 30x^3 + 55x^2 - 50x + 25$)

26. Work out:

a) $(x^3 + 2) \cdot [(4x^2 + 2) - (2x^2 + x + 1)]$ (Sol: $2x^5 - x^4 + x^3 + 4x^2 - 2x + 2$)

b) $(x^3 + 2) \cdot (4x^2 + 2) - (2x^2 + x + 1)$ (Sol: $4x^5 + 2x^3 + 6x^2 - x + 3$)

c) $(2x^2 + x - 2)(x^2 - 3x + 2) - (5x^3 - 3x^2 + 4)$ (Sol: $2x^4 - 5x^3 - x^2 + 8x - 4$)

d) $(x^2 - 3x + 2) \cdot [(5x^3 - 3x^2 + 4) - (2x^2 + x - 2)]$ (Sol: $5x^5 - 20x^4 + 12x^3 - x^2 - 20x + 12$)

e) $2x^2 + x - 2 - (x^2 - 3x + 2) \cdot (5x^3 - 3x^2 + 4)$ (Sol: $-5x^5 + 18x^4 - 19x^3 + 4x^2 + 13x - 10$)

4.4. Quotient of a polynomial by a monomial

Quotient of a polynomial by a monomial, (if possible) is another polynomial whose terms are obtained by dividing each term of the polynomial by the monomial.

Example 5: Divide $\dfrac{12x^3 - 9x^2 + 3x}{3x} = \dfrac{12x^3}{3x} - \dfrac{9x^2}{3x} + \dfrac{3x}{3x} = \boxed{4x^2 - 3x + 1}$

Exercises

27. Calculate the following quotients:

a) $\dfrac{6x^3 + 4x^2 - 8x}{2x}$

b) $\dfrac{4x^5 - 2x^4 + 4x^3 - 8x^2 + 16x}{-2x}$

c) $\dfrac{15x^5 - 10x^4 + 5x^2}{5x^2}$

d) $\dfrac{10x^7 - 12x^5 - 24x^3}{6x^3}$

Sol.: a) $3x^2 + 2x - 4$; b) $-2x^4 + x^3 - 2x^2 + 4x - 8$; c) $3x^3 - 2x^2 + 1$; d) $\dfrac{5}{3}x^4 - 2x^2 - 4$.

28. Calculate the following quotients:

a) $(8x^8 - 6x^4 - 4x^3) : (-4x^3)$

b) $\dfrac{-3x^4 + 6x^3 - 12x^2}{3x}$

c) $(-18x^3yz^3) : (6xyz^3)$

d) $\dfrac{-12x^9 + 2x^5 - x^4}{4x^4}$

Sol.: a) $-2x^5 + 3/2x + 1$; b) $-x^3 + 2x^2 - 4x$; c) $-3x^2$; d) $-3x^5 + 1/2x - 1$.

4.5. Factorization of expressions

It is important to know how to write expressions including brackets when it is possible; this is called **factorization**.

Example 6: Factorize: a) $12x + 6$ b) $21x - 14$ c) $10x^2 - 15x$

Solution:

a) $12x + 6 = \boxed{2} \cdot 2 \cdot \boxed{3} \cdot x + \boxed{2} \cdot \boxed{3} \cdot 1 =$ We mark the common factors $\boxed{2} \cdot \boxed{3} \cdot (2x + 1) = 6(2x + 1)$.

b) $21x - 14 = 3 \cdot \boxed{7} \cdot x - 2 \cdot \boxed{7} = 7(3x - 2)$.

c) $10x^2 - 15x = 2 \cdot \boxed{5} \cdot \boxed{x} \cdot x - 3 \cdot \boxed{5} \cdot \boxed{x} = 5x(2x - 3)$

We can check our answer by expanding the expression. *DO IT BY YOURSELF!!!*

Exercises

29. Factorize:

a) $10x + 15$ b) $4x - 12$ c) $9x - 9$ d) $3x^2 - 2x$ e) $9x^2 + 3x$

f) $3x - 3$ g) $33x^2 - 3x$ h) $5x - 3x^2$ i) $13x^2 - 2$

30. Factorize:

a) $x^2 + 3x$ b) $3x^2 - 5x$ c) $4x^4 + 7x^3 - 2x^2$ d) $35x^2 - 21$

e) $6x^2 + 15x$ f) $7x^2 - x$ g) $2x^2 - 4x + 2$ h) $- 6x^3 + 10x^2 + 2x$

31. Factorize, if possible:

a) $-5x^2 + 2x^3$ b) $3x^2 - 9x^3$ c) $3x^2 - 3x$ d) $x^3 - x^2$

e) $7x - 4y$ f) $3x^2 + 2$ g) $12x - 4y$ h) $5x^2 - 1$

32. Simplify the following algebraic fractions:

a) $\dfrac{5a + 5b}{5a + 10}$ b) $\dfrac{6x^3}{4x^2 + 2x^2}$ c) $\dfrac{x + x^2}{x^2 + x^3}$ d) $\dfrac{6x^3}{2x^3 + 4x^2}$

5. Polynomial or Remarkable identities

When working on polynomials, there are three identities that are going to be very useful for you to save time. They are named *polynomial identities, remarkable identities* or *special products*. They are:

Polynomial or Remarkable identities	
Binomial squares	**Sum x Difference:**
$(a+b)^2 = a^2 + 2ab + b^2$	
$(a-b)^2 = a^2 - 2ab + b^2$	$(a+b).(a-b) = a^2 - b^2$

These two figures may help you to memorize these identities. They also are useful as demonstrations.

$$(a+b)^2 = a^2 + 2ab + b^2 \qquad\qquad (a-b)^2 = a^2 - 2ab + b^2$$

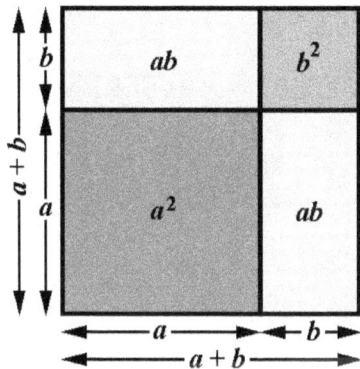   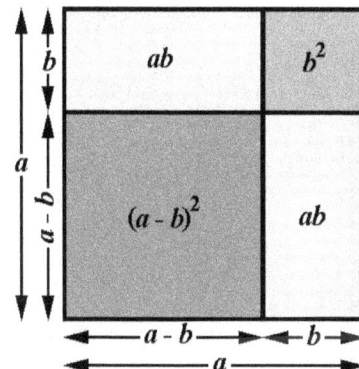

Squared of the sum of two monomials equals the square of the first monomial plus double of the product of both monomials plus square of the second one.

Squared of the difference of two monomials equals the square of the first monomial minus double of the product of both monomials plus square of the second one.

Example 7: Work out the following polynomial identities: a) $(x + 5)^2$ b) $(x - 6)^2$ c) $(x+2)\cdot(x-2)$

<u>Solution:</u>

a) $(x + 5)^2 = x^2 + 2\cdot x \cdot 5 + 5^2 = \boxed{x^2 + 10x + 25}$

b) $(3x - 4)^2 = (3x)^2 - 2\cdot 3x \cdot 4 + 4^2 = \boxed{9x^2 - 24x + 16}$

c) $(x + 2)\cdot(x - 2) = x^2 - 2^2 = \boxed{x^2 - 4}$.

---

33. Expand, by applying the SQUARE OF A SUM:

   a) $(x + 2)^2$ b) $(2x + 3)^2$ c) $(x + 1)^2$ d) $(x + y)^2$

   e) $(x + 3)^2$ f) $(x + 4)^2$ g) $(2x + 1)^2$

34. Expand, by applying the SQUARE OF A DIFFERENCE:

   a) $(x - 5)^2$ b) $(1 - 2x)^2$ c) $(x - 1)^2$ d) $(x - 4)^2$

   e) $(3x - 5)^2$ f) $(2a - 1)^2$ g) $(x - 3)^2$

35. Expand, by applying the A SUM MULTIPLIED BY A DIFFERENCE:

   a) $(x + 2) \cdot (x - 2)$ b) $(2x + 3) \cdot (2x - 3)$

   c) $(a + 1) \cdot (a - 1)$ d) $(x + 4) \cdot (x - 4)$

   e) $(2x + 1) \cdot (2x - 1)$ f) $(2a + 3b) \cdot (2a - 3b)$ g) $(x + 3) \cdot (x - 3)$

---

36. Work out the following special products:

   a) $(x + 2)^2$ (Sol: $x^2 + 4x + 4$)   b) $(x - 3)^2$ (Sol: $x^2 - 6x + 9$)

   c) $(x + 4)(x - 4)$ (Sol: $x^2 - 16$)   d) $(x + 3)^2$ (Sol: $x^2 + 6x + 9$)

   e) $(x - 4)^2$ (Sol: $x^2 - 8x + 16$)   f) $(x + 5)(x - 5)$ (Sol: $x^2 - 25$)

   g) $(a + 4)^2$ (Sol: $a^2 + 8a + 16$)   h) $(a - 2)^2$ (Sol: $a^2 - 4a + 4$)

   i) $(a + 3)(a - 3)$ (Sol: $a^2 - 9$)   j) $(2x + 3)^2$ (Sol: $4x^2 + 12x + 9$)

   k) $(3x - 2)^2$ (Sol: $9x^2 - 12x + 4$)   l) $(2x + 1)(2x - 1)$ (Sol: $4x^2 - 1$)

130

m) $(3x + 2)^2$ (Sol: $9x^2 +12x + 4$)   n) $(2x - 5)^2$ (Sol: $4x^2 - 20x+25$)

ñ) $(3x + 2)(3x - 2)$ (Sol: $9x^2 - 4$)   o) $(4b + 2)^2$ (Sol: $16b^2 + 16b + 4$)

p) $(5b - 3)^2$ (Sol: $25b^2 - 30b + 9$)   q) $(b + 1)(b - 1)$ (Sol: $b^2 -1$)

r) $(4a + 5)^2$ (Sol: $16a^2 + 40a + 25$)   s) $(5a - 2)^2$ (Sol: $25a^2 - 20a + 4$)

t) $(5a + 2)(5a - 2)$ (Sol: $25a^2 - 4$)   u) $(4y + 1)^2$ (Sol: $16y^2 + 8y + 1$)

v) $(2y - 3)^2$ (Sol: $4y^2 - 12y + 9$)   w) $(2y + 3)(2y - 3)$ (Sol: $4y^2 - 9$)

x) $(3x + 4)^2$ (Sol: $9x^2 +24x+16$)   y) $(3x - 1)^2$ (Sol: $9x^2 - 6x+1$)

z) $(3x + 4)(3x - 4)$ (Sol: $9x^2 - 16$)   a′) $(5b + 1)^2$ (Sol: $25b^2 +10b +1$)

b′) $(2x - 4)^2$ (Sol: $24x^2 -16x+16$)   c′) $(4x + 3)(4x - 3)$ (Sol: $16x^2 - 9$)

37. In this exercise, you must do the inverse operation. This is named ***factorizing***. Two first are made for you:

a) $x^2 - 25 = \boxed{(x + 5) \cdot (x - 5)}$ b) $x^2 - 18x + 81 = \boxed{(x - 9)^2}$

c) $x^2 - 9$ d) $x^2 - 20x + 100$ e) $x^2 - 81$ f) $x^2 - 14x + 49$

g) $x^2 - 100$ h) $x^2 + 22x + 121$ i) $x^2 - 49$ j) $x^2 + 18x + 81$

k) $9x^2 - 16$ l) $16x^2 - 9$ m) $x^2 - 36$ n) $25x^2 - 120x + 144$

ñ) $49x^2 + 84x + 36$ o) $x^2 - 16x + 64$ p) $81x^2 - 180x + 100$ q) $49x^2 - 64$

r) $x^2 - 1$ s) $25x^2 - 1$ t) $4x^2 - 12x + 9$ u) $x^2 - 6x + 9$

v) $x^2 + 10x + 25$ w) $x^2 - 121$

38. Reduce the following algebraic expressions to their simplest form:

a) $\dfrac{6x - 4}{9x^3 - 6x^2}$ b) $\dfrac{5x^2 + 10x}{x + 2}$ c) $\dfrac{x^3 + x^2}{2x^3 - 3x^2}$ d) $\dfrac{3x^3 - x^2}{x^3 + 2x^2}$ e) $\dfrac{x^2 + 2x + 1}{x^2 - 1}$ f) $\dfrac{x^2 - 4}{x^2 - 4x + 4}$

g) $\dfrac{2x^2 - 8}{x + 2}$ h) $\dfrac{2x + 1}{4x^2 + 4x + 1}$ i) $\dfrac{2x^4 - 2x^3}{4x^4 - 4x^2}$ j) $\dfrac{3x^4 - 9x^2}{x^2 - 3}$

Review exercises

1. If x and y are the ages today of two brothers, express the following statements by using two unknowns:

    a) The sum of their ages five years ago.

    b) The product of their ages in 6 years.

    c) Difference between the oldest and the youngest.

Sol.: a) $x + y - 10$; b) $xy + 6x + 6y + 36$; c) $x - y$.

2. Write the algebraic expression for the perimeter and area of these rectangles:

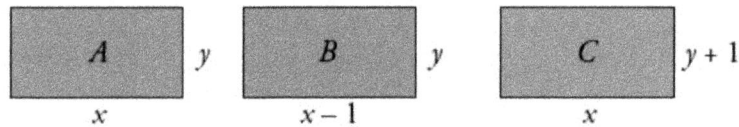

Sol.: a) $\begin{cases} P = 2x + 2y \\ A = x \cdot y \end{cases}$ b) $\begin{cases} P = 2x + 2y - 2 \\ A = x \cdot y - y \end{cases}$ c) $\begin{cases} P = 2x + 2y + 2 \\ A = x \cdot y + x \end{cases}$

3. Simplify:

    a) $6x^2 - 7x^2 + 3x^2$ b) $-6xy - 5xy + 10xy$ c) $\frac{1}{3}xy^2 - \frac{3}{5}xy^2 - \frac{7}{3}xy^2$

    d) $\frac{2x^3}{3} + \frac{1}{5}x^3 - x^3$ **Sol.:** a) $2x^2$; b) $-xy$; c) $-\frac{13}{5}xy^2$; d) $-\frac{2}{15}x^3$

4. Simplify:

    a) $5x - x^2 + 7x^2 - 9x + 2$ b) $2x + 7y - 3x + y - x^2$ c) $x^2y^2 - 3x^2y - 5xy^2 + x^2y + xy^2$

Sol.: a) $6x^2 - 4x + 2$; b) $-x^2 - x + 8y$; c) $x^2y^2 - 2x^2y - 4xy^2$

5. Work out the following products of monomials:

    a) $6x^2(-3x)$ b) $(2xy^2)(4x^2y)$ c) $\left(\frac{3}{4}x^3\right)\left(\frac{1}{2}x^3\right)$ d) $\left(\frac{1}{4}xy\right)\left(\frac{3xz}{2}\right)$

Sol.: a) $-18x^3$; b) $8x^3y^3$; c) $\frac{3}{8}x^6$; d) $\frac{3}{8}x^2yz$

6. Simplify the following expressions.

    a) $(2x^3 - 5x + 3) - (2x^3 - x^2 + 1)$ b) $5x - (3x + 8) - (2x^2 - 3x)$

    What is the degree of the resulting polynomials?

Sol.: a) $x^2 - 5x + 2$ (Degree 2); b) $-2x^2 + 5x - 8$ (Degree 2).

7. Simplify:

a) $(3x^3 - 5x^2 + x - 1) + (2x^4 + x^3 - 2x + 4)$

b) $(3x^3 - 5x^2 + x - 1) - (-x^3 + 3x^2 - 7x)$

c) $(3x^3 - 5x^2 + x - 1) - (2x^4 + x^3 - 2x + 4) + (-x^3 + 3x^2 - 7x)$

c) $(2a + 3b - 5ab) + (5a - 4b + 2ab) - (7a + b - ab)$

e) $(x^3 - 5x^2 + 3) - (2x^2 + 3x - 7) - (8x + 2)$

f) $(2x^2y - 3xy^2 + 5xy) - (6xy + 2x^2y - 3xy^2) + (5xy^2 - 3xy - 4x^2y)$

g) $(5x - 2) \cdot (x^3 - 4x^2 + 2x - 1)$

h) $(2a - 3b + 5) \cdot (6a - 4b + 2)$

i) $(-2x^2 - 4x + 3) \cdot (4x^2 - 3x - 6)$

j) $(5x - 3y + z) \cdot (2x + 4y - 9z)$

k) $(3x-2) \cdot (-5x+3) \cdot (x+4)$

l) $(2x-3)^3$

8. Work out, simplify and indicate the degree:

a) $x(x^2 - 5) - 3x^2(x + 2) - 7(x^2 + 1)$ b) $5x^2(-3x + 1) - x(2x - 3x^2) - 2 \cdot 3x$

Sol.: a) $-2x^3 - 13x^2 - 5x - 7$ (Degree 3);   b) $-12x^3 + 3x^2 - 6x$ (Degree 3).

9. Factorize:

a) $12x^3 - 8x^2 - 4x$ b) $-3x^3 + x - x^2$ c) $2xy^2 - 4x^2y + x^2y^2$

d) $\dfrac{2}{3}x^2 + \dfrac{1}{3}x^3 - \dfrac{5}{3}x$

Sol.: a) $4x(3x^2 - 2x - 1)$;   b) $x(-3x^2 + 1 - x)$;   c) $xy(2y - 4x + xy)$;   d) $\dfrac{1}{3}x(2x + x^2 - 5)$

10. Expand:

a) $(x + 6)^2$ b) $(7 - x)^2$ c) $(3x - 2)^2$

d) $\left(x + \dfrac{1}{2}\right)^2$ e) $(x - 2y)^2$ f) $\left(\dfrac{2}{5}x - \dfrac{1}{3}y\right)^2$

Sol.: a) $x^2 + 36 + 12x$;   b) $49 + x^2 - 14x$;   c) $9x^2 + 4 - 12x$;

d) $x^2 + \dfrac{1}{4} + x$;   e) $x^2 + 4y^2 - 4xy$;   f) $\dfrac{4x^2}{25} + \dfrac{y^2}{9} - \dfrac{4xy}{15}$.

133

11. Work out:

a) $(x + 7)(x - 7)$ b) $(3 + x)(3 - x)$ c) $(3 + 4x)(3 - 4x)$ d) $(x^2 + 1)(x^2 - 1)$

Sol.: a) $x^2 - 49$; b) $9 - x^2$; c) $9 - 16x^2$; d) $x^4 - 1$.

12. Express as a difference of squares:

a) $(2x + 7)(2x - 7)$ b) $(4x - 1)(4x + 1)$

c) $(x^2 + x)(x^2 - x)$ d) $(1 - 5x)(1 + 5x)$

Sol.: a) $4x^2 - 49$; b) $16x^2 - 1$; c) $x^4 - x^2$; d) $1 - 25x^2$.

13. Express as the square of a sum, the square of a difference or a difference of squares:

a) $x^2 + 9 - 6x$ b) $4x^2 + 1 + 4x$ c) $4x^2 - 9$

d) $9x^2 - 12x + 4$ e) $16x^2 - 1$ f) $16x^2 + 40x + 25$

Sol.: a) $(x - 3)^2$; b) $(2x + 1)^2$; c) $(2x + 3)(2x - 3)$; d) $(3x - 2)^2$; e) $(4x + 1)(4x - 1)$; f) $(4x + 5)^2$.

14. Simplify the following algebraic fractions:

a) $\dfrac{9x}{12x^2}$ b) $\dfrac{x(x+1)}{5(x+1)}$ c) $\dfrac{x^2(x+2)}{2x^3}$ **Sol.:** a) $\dfrac{3}{4x}$; b) $\dfrac{x}{5}$; c) $\dfrac{(x+2)}{2x}$

15. Simplify:

a) $\dfrac{x^2 - 4x}{x^2}$ b) $\dfrac{3x}{x^2 + 2x}$ c) $\dfrac{3x + 3}{(x+1)^2}$ d) $\dfrac{2x^2 + 4x}{x^3 + 2x^2}$

e) $\dfrac{8x^3 - 4x^2}{(2x-1)^2}$ f) $\dfrac{5x^3 + 5x}{x^4 + x^2}$ **Sol.:** a) $\dfrac{x-4}{x}$; b) $\dfrac{3}{x+2}$; c) $\dfrac{3}{(x+1)}$; d) $\dfrac{2}{x+2}$; e) $\dfrac{4x^2}{2x-1}$; f) $\dfrac{5}{x}$

16. JUST A CHALLENGE!!! Express as an algebraic expression the area of the coloured zone:

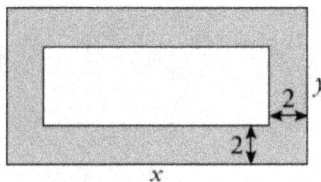

Sol.: 4x + 4y − 16.

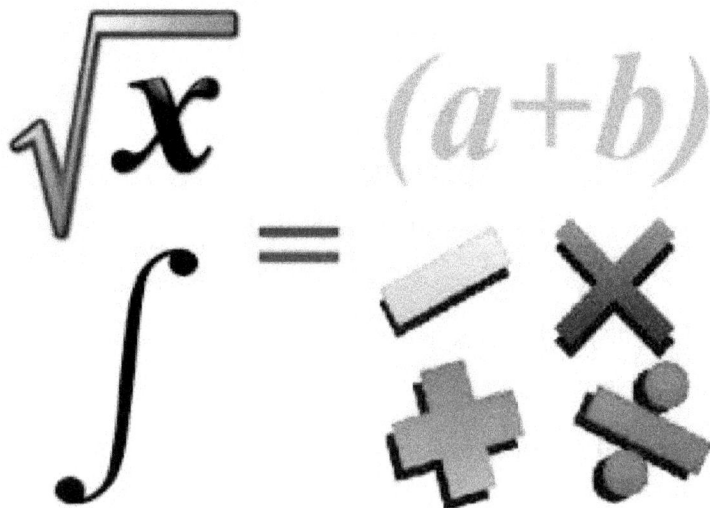

Unit 7.- Equations

1. Equations

An *equation* is a statement in which two expressions are equal.

The letter in an equation is called the *unknown*. (Sometimes it is called the *variable*).

These are examples of equations:

$$x = 2 \qquad 3 = x \qquad 2x = 4 \qquad 5x+1 = 3 \qquad x^2 = 4$$

In an equation we consider two *members*:

- First member is the member to the left of the sign "=".
- Second member is the member to the right of the sign "=".

Each monomial is called a term. Monomials with the same literal part are called *"like terms"*

Example:

In the equation $3t - 2 = -7$ we say:

- The first member is $3t - 2$
- The second member is -7
- There are three terms: $3t$; (-2) and (-7)
- The unknown is t.

In the equation $3x^2 - 6 = 5x + 2$, we say:

- The first member is $3x^2 - 6$
- The second member is $5x + 2$
- There are four terms: $3x^2$; (-6); $5x$ and 2
- The unknown is x.

1.1. Solution of an equation

The **solution** of an equation is a number that when substituted by the unknown, the equality is verified.

Example:

For the equation $x + 2 = 10$, the solution is $x = 8$, because $8 + 2 = 10$.

For the equation $3x - 5 = 1$, the solution is $x = 2$, because $3 \cdot 2 - 5 = 1 \rightarrow 6 - 5 = 1$.

For the equation $\dfrac{x-2}{3} = 5$, the solution is $x = 17$, because $\dfrac{17-2}{3} = 5 \rightarrow \dfrac{15}{3} = 5$.

Exercises

1. Check the given solutions for the following equations:

a) The solution for $5x + 2 = 22$ is $x = 4$

b) The solution for $\dfrac{x+5}{2} = 4$ is $x = 3$

2. Join each equation with its solution:

a) $2x - 1 = 5$ $x = 10$

b) $x - 8 = 2$ $x = (-3)$

c) $2 = x - 5$ $x = 3$

d) $4 + x = 4$ $x = 0$

e) $3 - 3x = 6$ $x = (-1)$

f) $x + 4 = 1$ $x = 7$

136

2. Solving equations

2.1. Starting with equations

Solving an equation is finding out its solution. This means that we must have the unknown alone in one of the members of the equation. We are seeing it with examples.

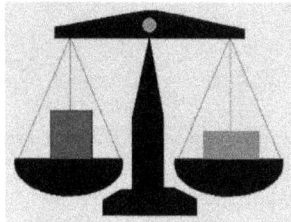

Figure 1

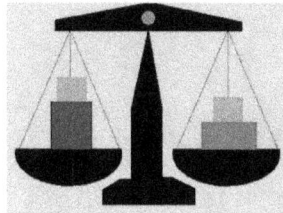

Figure 2

Look at the following equilibrated balance, in figure 1.

As you can see, you can add identical weights in both sides and the balance keeps equilibrated.

An equation can be considered as an equilibrated balance. So, we are going to be able to add or subtract identical *weights*.

Example: Solve the following equations:

a) x + 5 = 12

Solution: If we want to have the "x" alone, the number "5" must disappear. What would you do? If we subtract 5 in the first member, the number "5" will disappear. But if we subtract 5 in the first member, we must also subtract 5 in the second one:

$x + \cancel{5} - \cancel{5} = 12 - 5 \quad \rightarrow \quad x = 7$

CHECK: $x + 5 = 12 \rightarrow 7 + 5 = 12$ *OK!!!*

b) x – 2 = 4

Solution: If we want to have the "x" alone, the number "(- 2)" must disappear. What would you do? If we add 2 in the first member, the number "(-2)" will disappear. But if we add 2 in the first member, we must also add 2 in the second one:

$x - \cancel{2} + \cancel{2} = 4 + 2 \quad \rightarrow \quad x = 6$

CHECK: $x - 2 = 4 \rightarrow 6 - 2 = 4$ *OK!!!*

c) 3x = 12

Solution: If we want to have the "x" alone, the number "3" must disappear. What would you do? If we divide the first member by 3, the number "3" will disappear. But if we divide the first member by 3, we must also divide the second one by 3:

$$\frac{3x}{3} = \frac{12}{3} \quad \rightarrow \quad x = 4$$

CHECK: $3x = 12 \rightarrow 3 \cdot 4 = 12$ *OK!!!*

d) $\frac{x}{4} = 7$

Solution: If we want to have the "x" alone, the number "4" must disappear. What would you do? If we multiply the first member by 4, the number "4" will disappear. But if we multiply the first member by 4, we must also multiply the second one by 4:

$$\frac{x}{4} \cdot 4 = 7 \cdot 4 \quad \rightarrow \quad x = 28 \qquad\qquad \text{CHECK:} \quad \frac{x}{4} = 7 \rightarrow \frac{28}{4} = 7 \rightarrow \qquad\qquad OK!!!$$

e) 3 – x = 1

Solution: If we want to have the "x" alone, the number "3" must disappear. What would you do? If we subtract 3 in the first member, the number "3" will disappear. But if we subtract 3 in the first member, we must also subtract 3 in the second one:

$$3 - 3 - x = 1 - 3 \quad \rightarrow \quad -x = -2 \quad \rightarrow$$ But we do not want to have "- x", but "x". So, we must change the sign of the first member. Of course, we must also change the sign of the second member $\rightarrow$ x = 2.

CHECK: $3 - x = 1 \rightarrow 3 - 2 = 1$ *OK!!!*

f) – 3x = – 12

Solution: If we want to have the "x" alone, the number "(-3)" must disappear. *BE CAREFUL WITH THE SIGNS* What would you do? If we divide the first member by (-3), the number "(-3)" will disappear. But if we divide the first member by (-3), we must also divide the second one by (-3):

$$\frac{+3x}{+3} = \frac{-12}{-3} \quad \rightarrow \quad x = \frac{-12}{-3} = 4 \qquad \text{CHECK:} \quad -3x = -12 \rightarrow -3 \cdot 4 = -12 \qquad OK!!!$$

138

g) $5 = 2 - x$

Solution: Till now, we have left the "x" alone in the first member. But, if it is more comfortable, we can leave it in the second member. $5 = 2 - x$ In this case, if we want to have the "x" alone, the number "2" must disappear. What would you do? If we subtract 2 in the second member, the number "2" will disappear. But if we subtract 2 in the second member, we must also subtract 2 in the first one:

$5 - 2 = 2 - 2 - x \quad \rightarrow \quad 5 - 2 = -x \rightarrow 3 = -x$ But we do not want to have "- x", but "x". So, we must change the sign of the second member. Of course, we must also change the sign of the first member $\rightarrow -3 = x$.

CHECK: $5 = 2 - (-3) \rightarrow 5 = 2 + 3$ *OK!!!*

h) $8 = 2 - x$

Solution: Another option, if we do not like to have "-x", we can add "x" in both members.

$8 + x = 2 - x + x \quad \rightarrow \quad 8 + x = 2$

Now, we must subtract "8" in both members. $\quad 8 - 8 + x = 2 - 8 \quad \rightarrow \quad x = 2 - 8 = (-6)$

CHECK: $8 = 2 - x \quad \rightarrow \quad 8 = 2 - (-6) \quad \rightarrow \quad 8 = 2 + 6 = 8 \quad$ *OK!!!*

Exercises

3. Solve the following equations:

a) $x + 9 = 14$ b) $x + 3 = 6$ c) $-13 + x = 9$ d) $x - 7 = -9$

e) $-35 = x + 22$ f) $3x = -27$ g) $5x = 20$ h) $12 = 4x$

4. Solve the following equations:

a) $9x = 45$ b) $110 = 10x$ c) $3x = 33$ d) $50 = 10x$

e) $75 = 25x$ f) $6 - x = 1$ g) $735 - x = 600$ h) $2x = -10$

2.2. More difficult equations

In some cases, the equation has more than one term containing unknowns. For example,

$$4 + 2x = 6x - 10 + 2$$

Now, we are going to solve this kind of equations, step by step, with some examples:

Example : Solve the equation $4 + 2x = 6x - 10 + 2$

First of all, we must have together all the terms containing "x" in the same member. When you are not sure, take all the terms with "x" to the left member. You must do it as in the previous examples.

Step 1: "6x" term will disappear in the right member if we subtract "6x" in this member (and, of course, in the other one).

$$4 + 2x - 6x = 6x - 6x - 10 + 2$$
$$4 + 2x - 6x = -10 + 2$$
$$4 - 4x = -8$$

Step 2: Now, we are going to leave the term with "x" alone. We must remove the number "4". We must subtract 4 in both members.

$$4 - 4 - 4x = -8 - 4$$
$$-4x = -12$$

Step 3: Now, we are going to leave the "x" alone. We must divide both members by (-4).

$$\frac{-4x}{-4} = \frac{-12}{-4} \qquad \rightarrow \qquad x = \frac{-12}{-4} = \boxed{3}.$$

Step 4: CHECK:

$$4 + 2x = 6x - 10 + 2 \quad \rightarrow \quad x = 4$$
$$4 + 2 \cdot 3 = 6 \cdot 3 - 10 + 2$$
$$4 + 6 = 18 - 10 + 2$$
$$10 = 20 - 10 \qquad \qquad OK!!!$$

Example: Solve the equation $5 - 2x = 7x - 10 - 4x$

First of all, we must have together all the terms containing "x" in the same member. You are free to choose the member in which you take them. In this case, as there are more terms with "x" in the right member, we are going to take the "2x" to the right member.

Step 1: "(-2x)" term will disappear in the left member if we add "2x" in this member (and, of course, in the other one).

$$5 - 2\!\!\!/x + 2\!\!\!/x = 7x - 10 - 4x + 2x$$

$$5 = 7x - 10 - 4x + 2x \qquad \text{We add the "x" terms:}$$

$$5 = 5x - 10$$

Step 2: Now, we are going to leave the term with "x" alone. We must remove the number "(-10)". We must add 10 in both members.

$$5 + 10 = 5x \!\!\!/{+}\!\!\!/10 \!\!\!/{+}\!\!\!/10$$

$$15 = 5x$$

Step 3: Now, we are going to leave the "x" alone. We must divide both members by 5.

$$\frac{15}{5} = \frac{5\!\!\!/x}{5\!\!\!/} \qquad \rightarrow \qquad \frac{15}{5} = x = \boxed{3}.$$

Step 4: CHECK:
$$5 - 2x = 7x - 10 - 4x$$
$$5 - 2\cdot 3 = 7\cdot 3 - 10 - 4\cdot 3$$
$$5 - 6 = 21 - 10 - 12$$
$$- 1 = 21 - 22 \qquad\qquad OK!!!$$

Exercises	**5.** Solve the following equations: a) $3x + 2 = 14$ b) $3 - 2x = 5$ c) $5x + 12 = 2$ d) $3 = 4 - 3x$ e) $2x = x + 3$ f) $5x - 2 = x + 1$ g) $2x - 3 = 2x + 1$ h) $3x + 1 = 7x - 1$ **6.** Solve the following equations: a) $x + 8 + 2x = 6 - 2x$ b) $3 + 4x - 7 = x - 3$ c) $5x - 1 = 3x - 1 + 2x$ d) $6 - 3x + 2 = x + 7$ e) $2x + 5 - 3x = x + 19$ f) $7x - 2x = 2x + 1$

7. Solve the following equations:

a) $11 + 2x = 6x - 3 + 3x$ b) $7 + 5x - 2 = x - 3 + 2x$ c) $x - 1 - 4x = 5 - 3x - 6$

d) $5x = 4 - 3x + 5 - x$ e) $3x - x + 7x + 12 = 3x + 9$ f) $6x - 7 - 4x = 2x - 11 - 5x$

g) $7x + 3 - 8x = 2x + 4 - 6x$ h) $5x - 7 + 2x = 3x - 3 + 4x - 5$

8. Solve the following equations:

a) $7x - 6 = x + 8 + 5x$ b) $5x + 32 = 4x + 41$ c) $2 + 3x + 2x = 4x + 9$

d) $4x - 5 + x = 5 + 3x - 1$ e) $6x + 2 - 4x = 9 - x + 8$ f) $10 + x + 14 = 30 + 5$

9. Solve the following equations:

a) $18 + 2x - 8 = x - 25$ b) $8x - 6 = x + 8 + 6x$ c) $4x - 12 + x = 4x - 1$

d) $5x + 4 = 20 + 2x$ e) $4 - 6x = 3 + 4x$ f) $5x - 2 = 7x + 12$

10. Solve the following equations:

a) $3x + 5 = 2x - 6$ b) $2x - 3 + 5x = -2 - 6x + 3$ c) $3x - 1 = 5 - 3x$

3.3. Equations with parentheses

In some cases, the equations have one or more parentheses. For example, $10 + 2(x - 3) = 4x - 8$

In these cases, first of all, we must **expand** the parentheses.

Now, we are going to solve this kind of equations, step by step, with some examples:

Example: Solve the equation $10 + 2(x - 3) = 4x - 8$

Step 1: First of all, we must expand the parentheses: $+ 2(x - 3) = + 2x - 6$ And include it in the equation:

$$10 + 2x - 6 = 4x - 8$$

Step 2: We must have together all the terms containing "x" in the same member. For example, we are going to take "4x" to the left member. "4x" term will disappear in the right member if we subtract "4x" in this member (and, of course, in the other one).

$$10 + 2x - 6 - 4x = 4x - 4x - 8$$

$$10 + 2x - 6 - 4x = -8 \qquad \text{Notice that } \begin{cases} 10 - 6 = 4 \\ 2x - 4x = -2x \end{cases}$$

$$4 - 2x = -8$$

Step 3: Now, we are going to leave the term with "x" alone. We must remove the number "4". We must subtract 4 in both members.

$$4 - 4 - 2x = -8 - 4$$

$$-2x = -12$$

Step 4: Now, we are going to leave the "x" alone. We must divide both members by (-2).

$$\frac{-2x}{-2} = \frac{-12}{-2} \qquad \rightarrow \qquad x = \frac{-12}{-2} = \boxed{6}.$$

Step 5: CHECK

$$10 + 2(x - 3) = 4x - 8 \qquad \rightarrow \quad x = 6$$

$$10 + 2(6 - 3) = 4 \cdot 6 - 8$$

$$10 + 2(3) = 24 - 8$$

$$10 + 6 = 16 \qquad\qquad\qquad OK!!!$$

Example: Solve the equation $\quad 8 - 3(2 - 5x) = 4(x - 1)$

Step 1: First of all, we must expand the two parentheses: $\begin{cases} -3(2 - 5x) = -6 + 15x \\ 4(x - 1) = 4x - 4 \end{cases}$ And

include it in the equation:

$$8 - 6 + 15x = 4x - 4$$

Step 2: We must have together all the terms containing "x" in the same member. For example, we are going to take "4x" to the left member. "4x" term will disappear in the right member if we subtract "4x" in this member (and, of course, in the other one).

$$8 - 6 + 15x - 4x = 4x - 4x - 4$$

$$8 - 6 + 15x - 4x = -4 \qquad \text{Notice that } \begin{cases} 8 - 6 = 2 \\ 15x - 4x = 11x \end{cases}$$

$$2 + 11x = -4$$

Step 3: Now, we are going to leave the term with "x" alone. We must remove the number "2". We must subtract 2 in both members.

$$2 - 2 + 11x = -4 - 2$$

$$11x = -6$$

Step 4: Now, we are going to leave the "x" alone. We must divide both members by 11.

$$\frac{11x}{11} = \frac{-6}{11} \quad \rightarrow \quad x = \boxed{\frac{-6}{11}}$$

11. Solve the following equations:

a) $4 - (5x - 4) = 3x$

b) $7x + 10 = 5 - (2 - 6x)$

c) $5x - (4 - 2x) = 2 - 2x$

d) $1 - 6x = 4x - (3 - 2x)$

12. Find out the value of x:

a) $x - (3 - x) = 7 - (x - 2)$

b) $3x - (1 + 5x) = 9 - (2x + 7) - x$

c) $(2x - 5) - (5x + 1) = 8x - (2 + 7x)$

d) $9x + (x - 7) = (5x + 4) - (8 - 3x)$

13. Solve the following equations:

a) $2(x + 5) = 16$

b) $5 = 3 \cdot (1 - 2x)$

c) $5(x - 1) = 3x - 4$

d) $5x - 3 = 3 - 2(x - 4)$

e) $10x - (4x - 1) = 5 \cdot (x - 1) + 7$

14. Find out the value of x:

a) $6(x - 2) - x = 5(x - 1)$

b) $7(x - 1) - 4x - 4(x - 2) = 2$

c) $3(3x - 2) - 7x = 6(2x - 1) - 10x$

d) $4x + 2(x + 3) = 2(x + 2)$

15. Solve the following equations:

a) $3(x + 6) = -2(5 - x)$

b) $x + 9 = 2(x - 6)$

c) $2x + 3 = 4x + 6(x - 4) - 2$

d) $1 + 4(x - 2) = -3x + 5(x + 1)$

e) $2(x + 6) - 7x = 3x$

144

2.4. Equations with fractions

In some cases, the equations include fractions. For example, $\dfrac{x+2}{3} + \dfrac{5}{6} = 2x$

> In these cases, first of all, we must transform each term in a fraction being **equivalent** with the originals.

Now, we are going to solve this kind of equations, step by step, with some examples:

Example: Solve the equation $\dfrac{x+2}{3} + \dfrac{5}{2} = 2x$

Step 1: First of all, we must calculate the LCM (Lowest Common Multiple) of the denominators:

$$LCM(3, 2) = 6.$$

Step 2: Now, we are going to transform each term in a fraction being <u>equivalent</u> with the originals:

$\dfrac{x+2}{3} = \dfrac{}{6}$ Notice we have multiplied the denominator by 2. So, we also must multiply the

numerator by 2: $2 \cdot (x+2) = 2x+4$ So, we have $\dfrac{x+2}{3} = \dfrac{2x+4}{6}$

$\dfrac{5}{2} = \dfrac{}{6}$ Notice we have multiplied the denominator by 3. So, we also must multiply the

numerator by 3: $3 \cdot 5 = 15$ So, we have $\dfrac{5}{2} = \dfrac{15}{6}$

$\dfrac{2x}{1} = \dfrac{}{6}$ Notice we have multiplied the denominator by 6. So, we also must multiply the

numerator by 6: $6 \cdot 2x = 12x$ So, we have $\dfrac{2x}{1} = \dfrac{12x}{6}$

Step 3: Now, we change each term by its equivalent fraction:

$$\frac{2x+4}{6}+\frac{15}{6}=\frac{12x}{6}$$

Step 4: As in previous equations, we can multiply both members by the same number, in this case, by 6 (this means, we can remove all the denominators in an equation, when they are all similar):

$$2x + 4 + 15 = 12x \quad \rightarrow \quad 2x + 19 = 12x$$

Step 5: We must have together all the terms containing "x" in the same member. For example, we are going to take "12x" to the left member. "12x" term will disappear in the right member if we subtract "12x" in this member (and, of course, in the other one).

$$2x+19-12x = 12x-12x$$

$$-10x+19=0$$

Step 6: Now, we are going to leave the term with "x" alone. We must remove the number "19". We must subtract 19 in both members.

$$-10x+19-19=0-19$$

$$-10x=-19$$

Step 7: Now, we are going to leave the "x" alone. We must divide both members by (-10).

$$\frac{\cancel{-10}x}{\cancel{-10}}=\frac{-19}{-10} \quad \rightarrow \quad x=\frac{\cancel{-19}}{\cancel{-10}}=\boxed{\frac{19}{10}}$$

---

16. Solve the following equations:

a) $\dfrac{3x+3}{2}=2x-7$

b) $x+7=\dfrac{x+5}{3}$

c) $\dfrac{9x+3}{2}=\dfrac{2x+4}{3}$

d) $5-\dfrac{x+2}{3}=3-2x$

e) $2x+3=\dfrac{x+6}{3}$

f) $\dfrac{x+1}{3}-\dfrac{2x+2}{12}=1$

g) $5x-\dfrac{3-2x}{2}=2x-\dfrac{5}{2}$

h) $\dfrac{15x}{2}-\dfrac{x}{4}=\dfrac{5}{2}$

i) $\dfrac{4x-6}{5}=\dfrac{x}{3}-\dfrac{4}{15}$

j) $15x-\dfrac{6x-1}{2}-\dfrac{x-1}{3}=-6$

Exercises

17. Solve:

a) $\dfrac{10x-55}{2}=10x-\dfrac{95-10x}{2}$

b) $x+5=\dfrac{x+3}{3}$

c) $\dfrac{-4x+12}{4}=x-5$

d) $\dfrac{3+x}{2}=4$

e) $\dfrac{x}{2}+21=\dfrac{4x}{3}+24$

f) $\dfrac{5-9x}{8}+\dfrac{2x+3}{4}-\dfrac{143}{6}=2x$

g) $\dfrac{5x+7}{2}-\dfrac{3x+9}{4}=\dfrac{2x+4}{3}+5$

h) $1-\dfrac{x-5}{4}-\dfrac{x-3}{10}+\dfrac{x+3}{8}=0$

2.5. Steps to solve an equation

1.- Expand all **parentheses** in the equation.

2.- Eliminate **denominators** by multiplying every term by the LCM of the denominators.

3.- Move terms: Terms containing "x" must be together in one member, and non containing "x" terms, in the other one.

4.- Simplify **like terms** at both members, having an equation as **ax = b**.

5.- Calculate the value of the **unknown**.

Example: Solve the equation $\dfrac{x}{3}-\dfrac{3(x-2)}{5}=2\left(\dfrac{x-1}{3}\right)$

Solution:

Step 1: Parentheses: $\dfrac{x}{3}-\dfrac{3x-6}{5}=\dfrac{2x-2}{3}$

Step 2: Denominators: LCM (3,5, 3) = 15. We multiply each term by 15:

$5x - 3(3x - 6) = 5(2x - 2)$ We expand new parenthesis: $5x - 9x + 18 = 10x - 10$

Step 3: Move terms: $5x - 9x - 10x = -10 - 18$

Step 4: Simplify like terms: $-14x = -28$

Step 5: and, finally, $x=\dfrac{-28}{-14}=\boxed{2}$.

2.6. Problems with equations

18. The double of a number minus five is seventeen. What is the number?

19. The triple of a number plus four is forty-three. What is the number?

20. Martin thinks of a number, subtracts 15 and the answer is 14. What was the number?

21. Claudette thinks of a number, doubles it and adds 4. The answer is 25. What was the number?

22. Jhosir thinks of a number, multiplies it by 3 and subtracts 5. The answer is 25. What was the number?

23. Sean thinks of a number, halves it and adds 8. The answer is 20. What was the number?

24. The perimeter of this rectangle is 40 cm. Find *a*.

a

6 cm

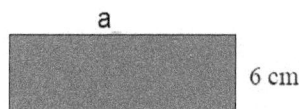

25. The area of this rectangle is 28 cm^2. Find d.

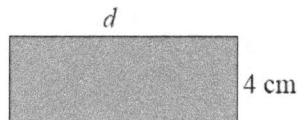

d

4 cm

26. The perimeter of this rectangle is 26 cm. Find n.

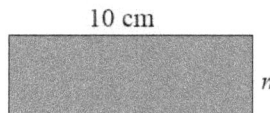

10 cm

n

27. The area of this rectangle is 20 cm^2. Find x.

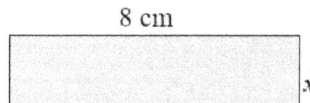

8 cm

x

28. The sum of my age and 7 is 42. Find my age.

29. The difference of a number and 23 is 124. Find the number.

30. The quotient of 35 and a number is 7. Find the number.

31. If someone gives me 24€ I will have 34.23 € Find the money I have.

32. The double of my age minus 7 years is the age of my elder brother who is 19 years old.

33. A teacher gives x coloured pencils each one of 8 girls except to one of them who only receives 5 pencils. The teacher gives 53 pencils in total. How many pencils did each girl receive?

34. Donald thinks of a number, multiplies it by 3 and subtracts 7. His answer is twice the number plus 5 units. Which is the number?

35. In a triangle, the smallest angle is 20° less than the median angle and the largest angle is twice the median one. Find all the angles.

36. If we shorten 2 cm each of the two opposite sides of a square, we get a rectangle with an area which has 28 cm² less than the area of the square. Find out the perimeter of the rectangle.

37. Solve these equations:

a) $\dfrac{3(x+2)}{2}+\dfrac{x-1}{5}=\dfrac{2(x+1)}{5}+\dfrac{37}{10}$

b) $\dfrac{2x-3}{2}-\dfrac{x+3}{4}=-4-\dfrac{x-1}{2}$

c) $\dfrac{1+12x}{4}+\dfrac{x-4}{2}=\dfrac{3(x+1)-(1-x)}{8}$

d) $\dfrac{3x-2}{6}-\dfrac{4x+1}{10}=-\dfrac{2}{15}-\dfrac{2(x-3)}{4}$

e) $\dfrac{2x-3}{6}-\dfrac{3(x-1)}{4}-\dfrac{2(3-x)}{6}+\dfrac{5}{8}=0$

f) $\dfrac{2}{3}(x+3)-\dfrac{1}{2}(x+1)=1-\dfrac{3}{4}(x+3)$

38. Solve these equations:

a) $3(x + 6) = -2 (5 - x)$

b) $1 + 4(x - 2) = -3x + 5(x + 1)$

c) $2(x + 3) -6(x + 5) = 3x + 4$

d) $3(5x + 9) - 3(x - 7) = 11 (x - 2)$

e) $\dfrac{10x-55}{2}=10x-\dfrac{95-10x}{2}$

f) $\dfrac{x+4}{5}-\dfrac{x+3}{4}=1-\dfrac{x+1}{2}$

g) $\dfrac{3x-7}{12}=\dfrac{2x-3}{6}-\dfrac{x-1}{8}$

h) $2+\dfrac{3x-1}{15}+\dfrac{x-4}{5}=\dfrac{x+4}{3}$

3. Quadratic equations

3.1. Quadratic equations

It is named *quadratic equation* an equation that can be transformed into another equivalent equation having this form:

$$ax^2 + bx + c = 0, \quad \text{with } a \neq 0$$

Solutions of a quadratic equation are the values of x that being substituted in it, make the equation to verify.

Quadratic equations are solved by using this expression: $\quad x = \dfrac{-b \pm \sqrt{b^2 - (4 \cdot a \cdot c)}}{2 \cdot a}$

Example: Solve: $x^2 - 3x + 2 = 0$

Solution:

$$\left. \begin{array}{l} a = 1 \\ b = (-3) \\ c = 2 \end{array} \right\} x = \frac{-(-3) \pm \sqrt{(-3)^2 - (4 \cdot 1 \cdot 2)}}{2 \cdot 1} = \frac{+3 \pm \sqrt{9-8}}{2} = \frac{+3 \pm 1}{2} \quad \rightarrow \quad \left\{ \begin{array}{l} x_1 = \dfrac{+3+1}{2} = \dfrac{4}{2} = 2 \\[2mm] x_2 = \dfrac{+3-1}{2} = \dfrac{2}{2} = 1 \end{array} \right.$$

So, solutions are $\boxed{2}$ and $\boxed{1}$.

Exercises

39. Solve these quadratic equations:

a) $x^2 - 6x + 8 = 0$ b) $x^2 - 8x + 7 = 0$ c) $4x^2 + 24x + 20 = 0$

d) $x^2 + 2x - 8 = 0$ e) $x^2 - 4x - 21 = 0$ f) $3x^2 + 6x - 12 = 0$

40. Solve these quadratic equations:

a) $2x^2 + 4x - 16 = 0$ b) $3x^2 - 6x - 9 = 0$ c) $x^2 + 9x + 20 = 0$

d) $x^2 + 4x + 4 = 16$ e) $x^2 - 6x + 9 = 25$ f) $4x^2 + 4x + 1 = 9$

3.2. Incomplete quadratic equations

In a quadratic equation $ax^2 + bx + c = 0$, parameters b or c, are zero, it is named as an *incomplete quadratic equation.*

Incomplete quadratic equations can be solved by using the general expression we have just studied, or by a shorter method:

· If $b = 0$, equation becomes $ax^2 + c = 0$. It is quickly solved.

· If $c = 0$, equation becomes $ax^2 + bx = 0$. It is solved by factorizing. **x(ax + b) = 0.**

Example: Solve: a) $3x^2 = 0$ b) $2x^2 - 8 = 0$ c) $x^2 - 12x = 0$

Solution:

a) $3x^2 = 0$ Dividing by 3: $x^2 = 0 \rightarrow x = \sqrt{0} = \boxed{0}$.

b) $2x^2 - 8 = 0$ Adding 8: $2x^2 = 8 \rightarrow$ Dividing by 2: $x^2 = 4 \rightarrow x = \sqrt{4} = \boxed{+2 \text{ and } (-2)}$.

c) $x^2 - 12x = 0$ Factorizing: $x(x - 12) = 0 \rightarrow \begin{cases} x = 0 \\ x - 12 = 0 \rightarrow x = 12 \end{cases}$

For this last step, you must remember if you have $a \cdot b = 0$, then, there are 2 possible solutions:

$\begin{cases} a = 0 \\ b = 0 \end{cases}$

41. Solve these incomplete quadratic equations:

a) $3x^2 - 12 = 0$ b) $5x^2 - 45 = 0$ c) $x^2 - 16 = 0$

d) $2x^2 - 6x = 0$ e) $6x^2 - 18x = 0$

42. Solve these incomplete quadratic equations:

a) $5x^2 - 20 = 0$ b) $3x^2 + 5x = 0$ c) $3x^2 - 75 = 0$ d) $5x^2 + 6x = 0$

e) $3x^2 = 0$ f) $x^2 - 16 = 0$ g) $4x^2 + 7x = 0$ h) $5x^2 - 15x = 0$

i) $2x^2 = 7x$ j) $3x^2 = 18x$

Exercises

43. Solve these quadratic equations:

a) $x^2 - 1 = 0$ b) $x^2 - 16 = 0$ c) $x(x+1) = 0$ d) $x(x-3) = 0$

e) $(x-1)^2 = 0$ f) $x^2 - 4 = 0$ g) $x^2 - 81 = 0$ h) $x(x-1) = 0$

i) $x(x+5) = 0$ j) $(x-3)^2 = 0$

44. Solve these incomplete quadratic equations, <u>without operating</u>:

a) $(3x+5) \cdot (2x-1) = 0$ b) $(6x-3) \cdot (2x-1) = 0$ c) $(2x-4) \cdot (x+6) = 0$

d) $(5x-8) \cdot (x-4) = 0$ e) $(x-4)(x-6) = 0$ f) $(x+2)(x-3) = 0$

g) $x(x+1)(x-5) = 0$ h) $(3x+1)(2x-3) = 0$

45. Solve these quadratic equations and check your solutions:

a) $x^2 + 2x = 0$ (0 y -2)   b) $2x^2 + 5x - 3 = 0$ (1/2 y -3)

c) $x^2 - 24 = 1$ (5 y -5)   d) $1 - 4x^2 = -8$ (3/2 y -3/2)

e) $x^2 - 6x = 0$ (0 y 6)   f) $3x^2 - 39x = 0$ (0 y 13)

3.3. Problems with quadratic equations

46. The area of a parallelogram is 50 cm^2. If the base is twice its height, calculate the height.

47. The breadth of a rectangular plot of land is 5 m less than its length. If the area of the plot is 104 m^2, find the dimensions of the plot.

48. A circle has an area of 154 cm^2. Find its radius.

49. In a triangle, its base is 3 cm less than its height. If its area is 14 cm^2, find its height.

50. The area of a rectangle is 165 cm^2. Find the length and the breadth of this rectangle if their difference is 4 cm.

A system of equations is formed by two equations. Each equation has two unknowns. For example:

$$\begin{cases} -6x + 5y = 26 \\ x - 2y = -9 \end{cases}$$

Systems of equations are also named *simultaneous equations*.

4.1. Solution of a system of equations

A pair of numbers *(x_0, y_0)*, are said to be *solution* of a *system of linear equations* with two unknowns

$$\begin{cases} ax + by = c \\ a'x + b\acute{y} = c' \end{cases}$$

if when substituting x by x_0 and y by y_0, both equalities are verified.

Example: Check if some of the following pairs ($x = -1, y = 4$) and ($x = 7, y = 8$) are solution of the following system of linear equations:

$$\begin{cases} -6x + 5y = 26 \\ x - 2y = -9 \end{cases}$$

Solution:

a) ($x = -1, y = 4$) $\begin{cases} -6(-1) + 5 \cdot 4 = 6 + 20 = 26 \rightarrow OK!!! \\ (-1) - 2 \cdot 4 = -1 - 8 = -9 \rightarrow OK!!! \end{cases}$

So, ($x = -1, y = 4$) is a solution of the system.

b) ($x = 7, y = 8$) $\begin{cases} -6 \cdot 8 + 5 \cdot 8 = -48 + 40 = -8 \neq 26 \rightarrow Not \ solution \\ 7 - 2 \cdot 8 = 7 - 16 = -9 \rightarrow OK!!! \end{cases}$

So, ($x = 7, y = 8$) is NOT a solution of the system, because it must be solution of <u>both</u> equations.

51. Check if some of the following pairs are solution of the following system of equations:

$$\begin{cases} -2x + 4y = 18 \\ 3x - 2y = 5 \end{cases}$$

a) $(x = -1, y = 4)$ b) $(x = 7, y = 8)$

52. Check if some of the following pairs are solution of the following system of equations:

$$\begin{cases} 5x + y = 43 \\ 3x + y = 1 \end{cases}$$

a) $(x = -1, y = 4)$ b) $(x = 7, y = 8)$

4.2. Graphic solution of a system of equations

Example: Represent the linear plot of equations $2x - y = 6$, $x + y = 0$. Find in the graph the solution of both equations. Check the solution by substituting in both equations.

Solution:

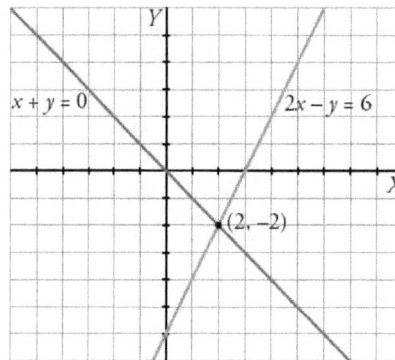

Common solution to both equations is $x = 2$, $y = -2$. Point $(2, -2)$.

Check your solution!!! $\begin{cases} 2x - y = 6 \rightarrow 2 \cdot 2 - (-2) = 4 + 2 = 6 \rightarrow OK!!! \\ x + y = 0 \rightarrow 2 + (-2) = 2 - 2 = 0 \rightarrow OK!!! \end{cases}$

53. Represent the following pairs of linear plots corresponding to each system of linear equations, find their solutions, and decide if they are equivalent.

a) $\begin{cases} x - 2y = -2 \\ 2x + 2y = 8 \end{cases}$

b) $\begin{cases} y - x = 0 \\ y = 2 \end{cases}$

4.3. Methods to solve a system of linear equations

There are three methods to solve systems of linear equations:

- Substitution method
- Equalization method
- Reduction method

4.3.1. Substitution method

We are going to describe **_substitution method_** with an example.

Example: Solve the following system by using the substitution method.

$$\begin{cases} x - 2y = -2 \\ 2x + 2y = 8 \end{cases}$$

Solution: You must follow the following steps:

Step 1: Isolate one unknown, where easier.	➡	In our case, easier unknown is "x" in the first equation: $x = -2 + 2y$
Step 2: Substitute this expression in the <u>other</u> equation.	➡	$2x + 2y = 8$. As $x = -2 + 2y$, then $2 \cdot (-2 + 2y) + 2y = 8$
Step 3: Solve this new equation with only one unknown.	➡	$2 \cdot (-2 + 2y) + 2y = 8$ $-4 + 4y + 2y = 8$ $4y + 2y = 4 + 8$ $6y = 12$ $y = \dfrac{12}{6} = 2$ So, $\boxed{y = 2}$.
Step 4: In order to obtain the value of the other unknown, we substitute known value in any of the two given equations. We choose it freely, but it is a good idea looking for the easiest one.	➡	$x = -2 + 2y$ As $y = 2$, then $x = -2 + 2 \cdot 2 = -2 + 4 = \boxed{2}$ So, solution is $\boxed{(x = 2, \ y = 2)}$.
Step 5: Check your solution.	➡	$\begin{cases} x - 2y = -2 \rightarrow 2 - 2 \cdot 2 = 2 - 4 = -2 \rightarrow OK!!! \\ 2x + 2y = 8 \rightarrow 2 \cdot 2 + 2 \cdot 2 = 4 + 4 = 8 \rightarrow OK!!! \end{cases}$

54. Solve the following systems by using the *substitution* method.

a) $\begin{cases} x + 3y = 5 \\ 5x + 7y = 13 \end{cases}$
b) $\begin{cases} 6x + 3y = 0 \\ 3x - y = 3 \end{cases}$
c) $\begin{cases} 3x + 9y = 4 \\ 2x + 3y = 1 \end{cases}$
d) $\begin{cases} x - 4y = 11 \\ 5x + 7y = 1 \end{cases}$

55. Solve the following systems by using the *substitution* method.

a) $\left. \begin{array}{l} 3x + 2y = 23 \\ x + y = 8 \end{array} \right\}$
b) $\left. \begin{array}{l} 3x + 2y = 14 \\ 2x - 5y = -16 \end{array} \right\}$
c) $\left. \begin{array}{l} x + 3y = 4 \\ 2x - y = 1 \end{array} \right\}$
d) $\left. \begin{array}{l} 3x - y = 17 \\ 2x + y = 8 \end{array} \right\}$

4.3.2. Equalization method

Example: Solve the following system by using the equalization method.

$$\begin{cases} 3x - 2y = 1 \\ 2x + 5y = 7 \end{cases}$$

Solution: You must follow the following steps:

Step 1: Isolate one unknown (choose freely) in both equations. As you are free, choose the easiest one.	⇨	We are going to isolate the "x": - First equation: $3x - 2y = 1 \rightarrow 3x = 2y + 1 \rightarrow \boxed{x = \dfrac{2y+1}{3}}$ - Second equation: $2x + 5y = 7 \rightarrow 2x = -5y + 7 \rightarrow \boxed{x = \dfrac{-5y+7}{2}}$
Step 2: Build an equality with both expressions.	⇨	$\dfrac{2y+1}{3} = \dfrac{-5y+7}{2}$
Step 3: Solve this new equation with only one unknown.	⇨	LCM(3,2) = 6, we must mutiply all the terms by 6. $\quad 2(2y+1) = 3(-5y+7)$ $\quad 4y + 2 = -15y + 21$ $\quad 4y + 15 = -2 + 21$ $\quad 19y = 19 \rightarrow y = \dfrac{19}{19} = 1$ So, $\boxed{y = 1}$.

Step 4: In order to obtain the value of the other unknown, we substitute known value in any of the two given equations. We choose it freely, but it is a good idea looking for the easiest one.	⟹	then $\qquad$ $2x + 5y = 7 \quad$ As $y = 1$, $2x + 5 \cdot 1 = 7$ $2x + 5 = 7$ $2x = 7 - 5 = 2$ $x = \dfrac{2}{2} = 1$ So, solution is $(x = 1, \ y = 1)$.
Step 5: Check your solution.	⟹	$\begin{cases} 3x - 2y = 1 \rightarrow 3 \cdot 1 - 2 \cdot 1 = 3 - 2 = 1 \rightarrow OK!!! \\ 2x + 5y = 7 \rightarrow 2 \cdot 1 + 5 \cdot 1 = 2 + 5 = 7 \rightarrow OK!!! \end{cases}$

Exercises

56. Solve the following systems by using the *equalization* method.

a) $\begin{cases} x + 3y = 5 \\ 5x + 7y = 13 \end{cases}$ b) $\begin{cases} 6x + 3y = 0 \\ 3x - y = 3 \end{cases}$ c) $\begin{cases} 3x + 9y = 4 \\ 2x + 3y = 1 \end{cases}$ d) $\begin{cases} x - 4y = 11 \\ 5x + 7y = 1 \end{cases}$

57. Solve the following systems by using the *equalization* method.

a) $\left.\begin{matrix} 7x - 5y = 52 \\ 2x + 5y = 47 \end{matrix}\right\}$ b) $\left.\begin{matrix} 2x - y = 5 \\ x + 2y = 5 \end{matrix}\right\}$ c) $\left.\begin{matrix} 3x - y = 7 \\ 2x + 3y = 1 \end{matrix}\right\}$ d) $\left.\begin{matrix} 7x + 4y = 3 \\ 5x - 3y = 8 \end{matrix}\right\}$

4.3.3. Reduction method

Reduction method is based on the fact that there are operations that can be done on a system of equations without changing its solution. These operations are named ***equivalence operations***, as they transform a system into another different one, but equivalent to original one:

- Adding both equations.
- Multiplying one equation by a real number and, after, adding.
- Multiplying both equations by real numbers and, after, adding.

Equivalence operation

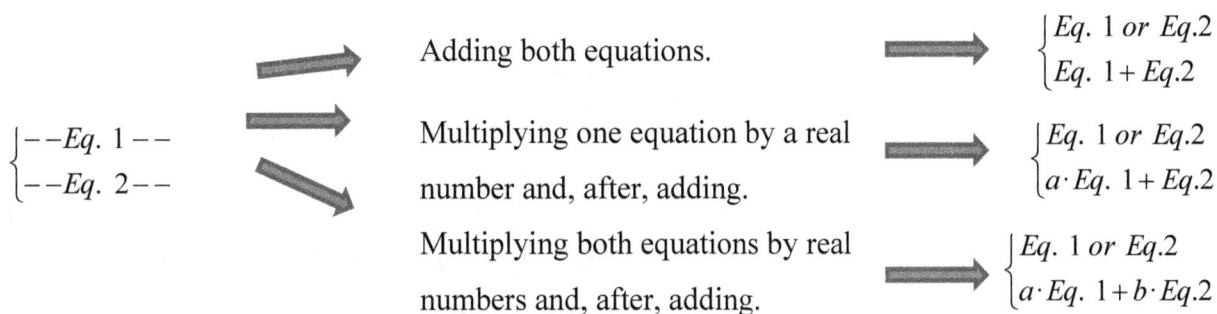

Adding both equations. → $\begin{cases} Eq.\ 1\ or\ Eq.2 \\ Eq.\ 1 + Eq.2 \end{cases}$

$\begin{cases} --Eq.\ 1\ -- \\ --Eq.\ 2\ -- \end{cases}$

Multiplying one equation by a real number and, after, adding. → $\begin{cases} Eq.\ 1\ or\ Eq.2 \\ a\cdot Eq.\ 1 + Eq.2 \end{cases}$

Multiplying both equations by real numbers and, after, adding. → $\begin{cases} Eq.\ 1\ or\ Eq.2 \\ a\cdot Eq.\ 1 + b\cdot Eq.2 \end{cases}$

We are going to describe *reduction method* with examples.

Example: Solve the following systems by using the reduction method:

a) $\begin{cases} 3x - 2y = 1 \\ 5x + 2y = 7 \end{cases}$ b) $\begin{cases} 3x + y = 1 \\ 5x + y = 3 \end{cases}$ c) $\begin{cases} 3x + 2y = 1 \\ 5x + y = 4 \end{cases}$ d) $\begin{cases} 3x + 5y = 1 \\ 2x + 3y = 1 \end{cases}$

Solution:

a) $\begin{cases} 3x - 2y = 1 \\ 5x + 2y = 7 \end{cases}$

Notice if you add both equations, terms containing unknown *"y"* will be cancellated.

$$\begin{cases} 3x - 2y = 1 \\ 5x + 2y = 7 \end{cases}$$

Addition: 8x   = 8 Solving this equation is easy, being x = $\boxed{1}$.

In order to obtain the value of the other unknown, we substitute known value in any of the two given equations. We choose it freely, but it is a good idea looking for the easiest one.

5x + 2y = 7   As x = 1, then

5·1 + 2y = 7

5 + 2y = 7

2y = 7 – 5 = 2

$x = \dfrac{2}{2} = 1$

So, solution is $\boxed{(x = 1,\ y = 1)}$.

Finally, check your solution.

$\begin{cases} 3x - 2y = 1 \rightarrow 3\cdot 1 - 2\cdot 1 = 3 - 2 = 1 \rightarrow OK!!! \\ 5x + 2y = 7 \rightarrow 5\cdot 1 + 2\cdot 1 = 5 + 2 = 7 \rightarrow OK!!! \end{cases}$

b) $\begin{cases} 3x + y = 1 \\ 5x + y = 3 \end{cases}$

Notice if you add both equations, none of the unknowns are being cancelated. We need a (-)

Sign before any of the *"y"* unknowns. So, we must multiply one of the equations by (-1).

$$\begin{cases} 3x + y = 1 \text{ \underline{\textit{Multiply by}} (-1)} \\ 5x + y = 3 \text{ \underline{\textit{Leave}}} \end{cases} \qquad \begin{cases} -3x - y = -1 \\ 5x + y = 3 \end{cases}$$

Now, if you add both equations, terms containing unknown *"y"* will be canceled.

$$\begin{cases} -3x - y = -1 \\ 5x + y = 3 \end{cases}$$

Addition: $2x = 2$ Solving this equation is easy, being x = $\boxed{1}$.

In order to obtain the value of the other unknown, we substitute known value in any of the two

Given equations. We choose it freely, but it is a good idea looking for the easiest one.

$5x + y = 3$ As x = 1, then

$5 \cdot 1 + y = 3$

$5 + y = 3$

$y = 3 - 5 = \boxed{(-2)}$

So, solution is $\boxed{(x = 1, \ y = (-2))}$.

Finally, check your solution. $\begin{cases} 3x + y = 1 \rightarrow 3 \cdot 1 + (-2) = 3 - 2 = 1 \rightarrow OK!!! \\ 5x + y = 3 \rightarrow 5 \cdot 1 + (-2) = 5 - 2 = 3 \rightarrow OK!!! \end{cases}$

c) $\begin{cases} 3x + 2y = 1 \\ 5x + y = 4 \end{cases}$

Notice if you add both equations, none of the unknowns are being canceled. We need a (-2) before *"y"* unknown in second equation. So, we must multiply second equation by (-2).

$$\begin{cases} 3x + 2y = 1 \ \underline{Leave} \\ 5x + y = 4 \ \underline{Multiply\ by\ (-2)} \end{cases} \longrightarrow \begin{cases} 3x + 2y = 1 \\ -10x - 2y = -8 \end{cases}$$

Now, if you add both equations, terms containing unknown *"y"* will be canceled.

$$\begin{cases} 3x + 2y = 1 \\ -10x - 2y = -8 \end{cases}$$

Addition: -7x = -7 Solving this equation is easy, being x = $\boxed{1}$.

In order to obtain the value of the other unknown, we substitute known value in any of the two given equations. We choose it freely, but it is a good idea looking for the easiest one.

5x + y = 4   As x = 1, then

5·1 + y = 4

5 + y = 4

y = 4 – 5 = $\boxed{(-1)}$

So, solution is $\boxed{(x = 1, \ y = (-1))}$.

Finally, check your solution. $\begin{cases} 3x + 2y = 1 \rightarrow 3·1 + 2·(-1) = 3 - 2 = 1 \rightarrow OK!!! \\ 5x + y = 4 \rightarrow 5·1 + (-1) = 5 - 1 = 4 \rightarrow OK!!! \end{cases}$

d) $\begin{cases} 3x+5y=1 \\ 2x+3y=1 \end{cases}$

Notice if you add both equations, none of the unknowns are being cancelled. Moreover, even multiplying any of the equations by a real number, we would not be able to cancel any unknown.

When this happens, you must choose one of the unknowns and *cross-multiplying*.

For example, if we work on "x", we can multiply by 2 above and by (-3) below:

$\begin{cases} 3x+5y=1 \ \underline{\textit{Multiply by 2}} \\ 2x+3y=1 \ \underline{\textit{Multiply by (–3)}} \end{cases}$ $\quad \begin{cases} 6x+10y=2 \\ -6x-9y=-3 \end{cases}$

Now, if you add both equations, terms containing unknown "x" will be cancelled.

$$\begin{cases} 6x+10y=2 \\ -6x-9y=-3 \end{cases}$$

Addition: $y=-1$ So, $y=\boxed{(-1)}$.

In order to obtain the value of the other unknown, we substitute known value in any of the two given

equations. We choose it freely, but it is a good idea looking for the easiest one.

   $2x + 3y = 1$ As $x = 1$, then

$2x + 3 \cdot (-1) = 1$

   $2x - 3 = 1$

   $2x = 1 + 3 = 4 \rightarrow 2x = 4 \rightarrow x = \boxed{2}$

So, solution is $\boxed{(x = 2, \ y = (-1))}$.

Finally, check your solution. $\begin{cases} 3x+5y=1 \rightarrow 3\cdot2+5\cdot(-1)=6-5=1 \rightarrow OK!!! \\ 2x+3y=1 \rightarrow 2\cdot2+3\cdot(-1)=4-3=1 \rightarrow OK!!! \end{cases}$

58. Solve the following systems by using the *reduction* method.

a) $\begin{cases} 3x + 5y = 11 \\ 4x - 5y = 38 \end{cases}$
b) $\begin{cases} x + 3y = 5 \\ 5x + 7y = 13 \end{cases}$
c) $\begin{cases} x - 4y = 11 \\ 5x + 7y = 1 \end{cases}$
d) $\begin{cases} 6x + 3y = 0 \\ 3x - y = 3 \end{cases}$

59. Solve the following systems by using the *reduction* method.

a) $\left.\begin{matrix} 7x - 5y = 52 \\ 2x + 5y = 47 \end{matrix}\right\}$
b) $\left.\begin{matrix} 2x - y = 5 \\ x + 2y = 5 \end{matrix}\right\}$
c) $\left.\begin{matrix} 3x - y = 7 \\ 2x + 3y = 1 \end{matrix}\right\}$
d) $\left.\begin{matrix} 7x + 4y = 3 \\ 5x - 3y = 8 \end{matrix}\right\}$

60. Solve the following systems by any method:

a) $\left.\begin{matrix} x + y = 2 \\ x - y = 6 \end{matrix}\right\}$
b) $\left.\begin{matrix} x + y = 12 \\ x - y = 2 \end{matrix}\right\}$
c) $\left.\begin{matrix} x + y = 5 \\ -x + y = 1 \end{matrix}\right\}$
d) $\left.\begin{matrix} 3x - 4y = -6 \\ x + 2y = 8 \end{matrix}\right\}$

c) $\left.\begin{matrix} 3x - y = 17 \\ 2x + y = 8 \end{matrix}\right\}$
d) $\left.\begin{matrix} x + 2y = 7 \\ 7x - 4y = 13 \end{matrix}\right\}$

4.4. Problems with systems of equations

Although lots of problems in algebra can be solved by using an equation with an only unknown, they become much easier if solved with a system of equations with two unknowns. We are seeing some examples.

Example: If 4 apples and 2 oranges cost 1 € and 2 apples and 3 oranges equals 0.70 €, how much does each apple and each orange cost?

Solution:

Setting	System of equation	Solution
x = price of one apple.	$\begin{cases} 4x + 2y = 1 \\ 2x + 3y = 0.70 \end{cases}$	Feel free to choose one of the methods.
y = price of one orange.		Remember to check your solution.
		- x = one apple costs 0.2 €
		- y = one orange costs 0.1 €.

Example: Find the value of two numbers if their sum is 12 and their difference is 4.

Solution:

Setting	System of equation	Solution
x = highest number.	$\begin{cases} x + y = 12 \\ x - y = 4 \end{cases}$	Feel free to choose one of the methods.
y = lowest number.		Remember to check your solution.
		Our numbers are 8 and 4.

60. The sum of the digits of a certain two-digit number is 7. Reversing its digits, increases the number by 9. What is the number?

61. The school that Stefan goes to is selling tickets to a choral performance. On the first day of tickets sales the school sold 3 senior tickets and 1 child ticket for a total of $38. The school took in $52 on the second day by selling 3 senior tickets and 2 child tickets. Find the price of a senior ticket And a child ticket.

62. Harry pays £8.50 for 5 kg of flour and 3 kg of sugar. Sarah pays £13.20 for 8 kg of flour and 4 kg of sugar. If the cost of flour is £f per kg and the cost of sugar is £ s per kg, write down two equations in s and f and solve them.

63. John and David have £ 14.00 altogether. If John's money is doubled and David's tripled, the will have £ 34.00 altogether. How much does each boy have?

64. A retailer can buy either two television sets and three video-recorders for £ 3750, or four television sets and one video-recorder for £4250. What is the cost of a television set? What is the cost of a video-recorder?

65. A toothbrush and a tube of toothpaste cost £ 4.15; the toothbrust costs 25 p less than the tube of toothpaste. Find the cost of each item.

66. Mrs Rogers bought 3 blouses and 2 scarves. She paid £ 26. Mrs Summers bought 4 blouses and 1 scarf. She paid £ 28. Find the cost of one blouse and one scarf.

67. Denise sells 300 tickets for a concert. Some tickets are sold to adults at £5 each and some are sold to children at £ 4 each. If she collects £ 1444 in tickets sales, how many tickets have been sold to adults and how many to children?

Exercises

68. A machine sells tickets for travel on a tram system. A single ticket costs £ 1 and a return ticket costs £ 2. In one day, the machine sells 100 tickets and takes £ 172. How many of each type of ticket were sold?

69. A bus stops at two places. The fare to the first stop is £ 3 and the fare to the second stop is £ 5. When the bus sets out, there are 38 people on the bus and the driver has collected £ 158 fares. How many people get off the bus at the first stop?

70. Find two numbers knowing that the difference between them is 8, and the difference between three times the lowest and the highest is 16.

71. Two integers can be added together to make 9, and the difference between twice the first one and three times the second one is 48. What are these numbers?

72. Find two numbers which can be added together to make 37 and knowing that the difference between them is 13.

73. The difference between the two acute angles of a right-angled triangle is 15°. What is the measure of the angles?

Review exercises

Linear equations

1. Solve the following linear equations:

a) $4x - 3 + 3x = 5x + 7$

b) $3(x - 2) + 5x - 6 = 6x + 2$

c) $3x - 4 - 5x = 2(x + 1) + 6$

d) $x - 6 + 2(x - 3) = 2x + 1 - 3(x - 1)$

e) $6x - 1 + 5(1 - x) + x = 4(2x - 1)$

f) $5x + 2(3x - 1 - x) = 9 + x - 14$

g) $2(x - 5) - 3(2x - 1) + 3 = -(x + 2) - 5$

Sol.: a) $x = 5$; b) $x = 7$; c) $x = (-3)$; d) $x = 4$; e) $x = 4/3$; f) $x = (-3/8)$; g) 1.

2. Solve the following linear equations:

a) $\dfrac{x}{3} - \dfrac{x}{4} + \dfrac{5}{6} = 1$

b) $\dfrac{x}{2} + \dfrac{x}{4} + \dfrac{x}{8} = \dfrac{21}{8}$

c) $\dfrac{x}{5} - \dfrac{2x}{3} = \dfrac{x}{10} - \dfrac{17}{30}$

d) $-\dfrac{1}{2} + \dfrac{x-2}{3} - \dfrac{x}{2} = -2$

e) $\dfrac{2+x}{3} - \dfrac{x}{7} = 2$

f) $\dfrac{6+x}{4} = \dfrac{x-2}{2} + 3$

g) $x - \dfrac{x}{2} = \dfrac{3}{8} - \dfrac{x}{4}$

h) $\dfrac{x}{5} - x = -\dfrac{2}{15} - \dfrac{3x}{5}$

i) $\dfrac{x}{2} - \dfrac{1}{6} - x = \dfrac{7}{2} + \dfrac{1-3x}{3}$

j) $\dfrac{x-1}{2} + \dfrac{2x-1}{9} = \dfrac{x}{6}$

Sol.: a) x=2; b) x=3; c) x=1; d) x=5; e) x=7; f) x=(-2); g) x=1/2; h) x=2/3; i) x=8; j) x=11/10.

3. A number and the number before it add up to 77. What numbers are they?   **Sol.:** 38 and 39.

4. When we add 13 units to the double of a number, the result is 99. What is the number?

Sol.: 43.

5. A pride of lions is made up of 13 lions, and there are 3 females more than males. How many lions and how many lionesses are there in the pride?   **Sol.:** 5 lions and 13 lionesses.

6. There are 31 people in a bar. How many men and women are there if we know that there are 5 more men than women?   **Sol.:** 13 women and 18 men.

7. A blouse has x buttons. A shirt has x+2 buttons.

    a) Write an expression for the number of buttons on four blouses.

    b) Write an expression for the number of buttons on three shirts.

    c) Three shirts have the same number of buttons in total as four blouses. Write an equation and solve it to find the value of x.

    d) How many buttons are there on a shirt?

Sol.: a) 4x; b) 3x + 6; c) 3x + 6 = 4x; d) 8 buttons.

8. Arthur has a bag containing 28 candies, some of them are mints and some others are lemon. If the number of mints is three times the number of lemon ones, how many candies of each kind can Arthur eat?   **Sol.:** 7 lemons and 21 mints.

9. Rachel and Sarah have 45 biscuits between them. Sarah tells Rachel: "if you give me 5 biscuits I will have twice as many biscuits as you". How many biscuits does each of them have?

Sol.: Rachel had 20 biscuits and Sarah had 25 biscuits.

10. Find the dimensions of a rectangular lawn knowing that the perimeter is 180 m long, and one side is 20 m longer than the other. **Sol.:** 35 m x 55 m.

11. A father is 25 years older than his son. Find their ages taking into account that in 15 years the father's age will be twice his son's age. **Sol.:** Son: 10; father: 35.

12. Find the ages of a mother and her daughter taking into account that three years ago the mother was three times older than the daughter, and in 8 years the mother's age will be twice the daughter's age. **Sol.:** Today they are 36 and 14.

Quadratic equations

13. Solve these quadratic equations and check your solutions:

a) $3x^2 - 16 = 2x^2$ (4 y -4)   b) $-x^2 + 9x = 0$ (0 y 9)

c) $x^2 - 7x - 18 = 0$ (-2 y 9)   d) $7x^2 + 21x - 28 = 0$ (1 y -4)

e) $x^2 - 10x + 9 = 0$ (9 y 1)   f) $x^2 - 26x + 25 = 0$ (25 y 1)

g) $4x^2 - 37x + 9 = 0$ (9 y ¼)   h) $4x^2 - 17x + 4 = 0$ (4 y ¼)

i) $x^2 - 25x + 144 = 0$ (9 y 16)   j) $4x^2 - 9x + 2 = 0$ (2 y ¼)

k) $2x^2 - 5x + 3 = 0$ (1,3/2)

14. When 2 is added to a certain number, the result is the same as dividing 8 by the number. Solve this problem finding the number.

Systems of equations

15. Check if x = 2 and y = (-1) is a solution for the following systems of equations:

a) $\begin{cases} 2x - y = -4 \\ 5x + y = -10 \end{cases}$ b) $\begin{cases} 3x - 4y = 10 \\ 4x + 3y = 5 \end{cases}$

Sol.: a) It is not solution; b) It is solution.

16. Represent the plots and find the solution of the following systems of equations:

a) $\begin{cases} 3x - y = 1 \\ x + 2y = 5 \end{cases}$ b) $\begin{cases} 3x - y = 0 \\ 3x + y = -6 \end{cases}$ c) $\begin{cases} x + 3y = -5 \\ 2x - y = 4 \end{cases}$ d) $\begin{cases} 2x - 3y = -4 \\ x + 8y = -2 \end{cases}$

Sol.: a) (1, 2); b) (-1, -3); c) (1, -2); d) (-2, 0).

17. Solve by using the substitution method:

a) $\begin{cases} x + 3y = 0 \\ 2x + y = -5 \end{cases}$ b) $\begin{cases} 8x - 3y = -25 \\ x - 5y = -17 \end{cases}$ c) $\begin{cases} 7x - y = -6 \\ 4x + 3y = 3 \end{cases}$ d) $\begin{cases} 2x + 16 = 2y \\ 2y - 3x = 16 \end{cases}$

Sol.: a) (-3, 1); b) (-2, 3); c) (-3/5, 9/5); d) (0, 8).

18. Solve by using the equalization method:

a) $\begin{cases} x = 4 \\ x - y = 6 \end{cases}$ b) $\begin{cases} x + 3y = -4 \\ x - 2y = 6 \end{cases}$ c) $\begin{cases} y = 6x \\ 7x = 2y - 5 \end{cases}$ d) $\begin{cases} 3x - 4y = -4 \\ 2x + y = -1 \end{cases}$

Sol.: a) (4, -2); b) (2, -2); c) (1, 6); d) (-8/11, 5/11).

19. Solve by using the reduction method:

a) $\begin{cases} x + y = 0 \\ x - y = 2 \end{cases}$ b) $\begin{cases} 3x - y = 0 \\ 3x + y = -6 \end{cases}$ c) $\begin{cases} 4x - 3y = 2 \\ 2x + y = -4 \end{cases}$ d) $\begin{cases} x + 2y = 1 \\ 3x - y = 7 \end{cases}$

e) $\begin{cases} x - 3y = 1 \\ 3x + 6y = 2 \end{cases}$ f) $\begin{cases} 3x + 2y = 3 \\ x + y = 7/6 \end{cases}$

Sol.: a) (1, -1); b) (-1, -3); c) (-1, -2); d) (15/7, -4/7); e) (4/5, -1/15); f) (2/3, 1/2).

21. Solve the following systems by using the method you consider more suitable:

a) $\begin{cases} x - y = 1 \\ 4x - 3y = 8 \end{cases}$ b) $\begin{cases} 3x = 1 + y \\ 3 + 2y = 10x \end{cases}$ c) $\begin{cases} 2x + 5y = -1 \\ 4x - 3y = -2 \end{cases}$ d) $\begin{cases} 3x - 2y = 2 \\ x + 4y = -5/3 \end{cases}$

Sol.: a) (5, 4); b) (1/4, -1/4); c) (-1/2, 0); d) (1/3, -1/2).

22. Find out two numbers whose sum is 160 and difference is 34.　　　　　**Sol.:** 97 and 63.

23. Two pens and three notebooks cost 7.80 €; five pens and four notebooks cost 13.2 €. What are the prices of each product?　　　　　**Sol.:** Pen: 1.2 €; Notebook: 1.8 €.

24. In a book shop, they have sold 45 books, some of them cost 32 € each, and some others, 28 € each. They got 1368 €. How many of each kind of books were sold?

Sol.:　27 books cost 32 € and 18 books cost 28 €.

25. In a farm, there are rabbits and chickens, having a total quantity of 29 heads and 92 legs. How many rabbits and chickens are there? **Sol.:** 12 chickens and 17 rabbits.

26. A test exam has 50 questions and I must answer all of them. Each correct answer adds 1 point for me and each wrong answer, will subtract 0.5 points. If mi mark has been 24.5 points, how many correct and wrong answers have I had?

Sol.: 33 correct and 17 wrong answers.

27. An olive oil company has bottled 2000 oil liters into bottles of 1.5 L and 2 L. If they have used 1100 bottles, how many of them are 1.5 L and 2 L? **Sol.:** 400 and 700.

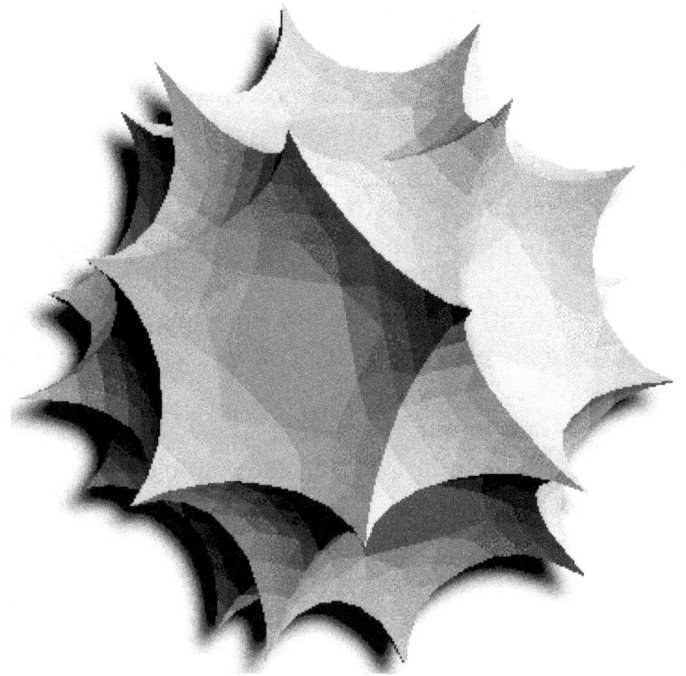

Unit 8.- Geometric elements

1. Straight lines

As you already know, infinite straight lines are passing by a point, while only one straight line is able to pass by two points.

Infinite straight lines are passing by a point	An only straight line is able to pass by two points

Relative positions of two straight lines

Two straight lines can be classified, according to their relative position, into *secant, parallel* or *coincident* straight lines.

> **- Secant:** Two straight lines are said to be secant if they intersect in a point. They have an only point in common. They have different slopes.
>
> **- Parallel:** Two straight lines are said to be parallel if they do not intersect themselves. They do not have any point in common. They have identical slopes.
>
> **- Coincident:** Two straight lines are said to be coincident if they pass by the same points. Their plots, together, are an only straight line. So, they have infinite points in common. They have identical slopes.

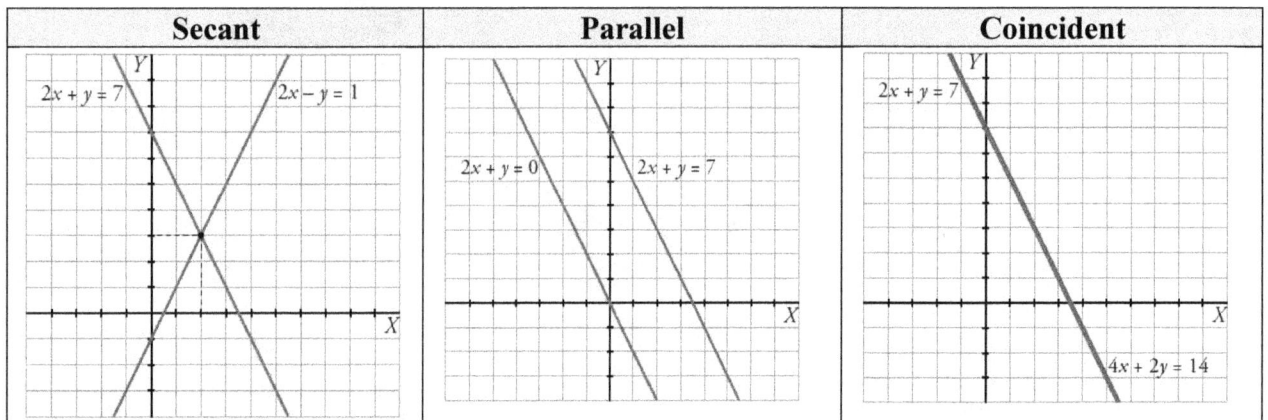

Secant	Parallel	Coincident
$2x + y = 7$ $2x - y = 1$	$2x + y = 0$ $2x + y = 7$	$2x + y = 7$ $4x + 2y = 14$

Moreover, *secant* straight lines can be classified, depending on their angle of intersection, into **perpendicular** (angle = 90º) or **non-perpendicular** (angle ≠ 90º).

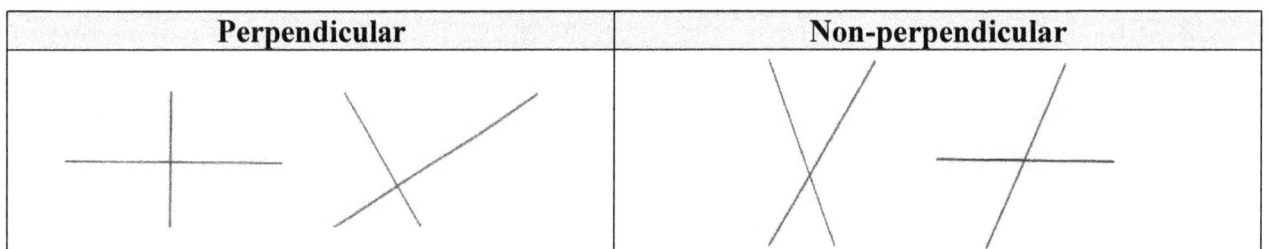

Perpendicular	Non-perpendicular

Semi-straight lines, segment, bisector

These are three important concepts.

- Semi-straight line: While a straight line is infinite (it does not finish in any of its heads), a semi-straight line finishes in one of its heads. A straight line can be divided into two semi-straight lines.

- Segment: Segment is a portion of straight line that has two extremes.

- Bisector of a segment: It is the straight line that divides a segment into two identical portions and, moreover, is perpendicular to the segment.

Semi-straight line	Segment	Bisector of a segment

2. Angles

An *angle* is a shape, formed by two lines or rays diverging from a common point (the *vertex*).

Parts of an angle:
  Vertex The vertex is the common point at which the two lines or rays are joined.
  Legs   The legs (sides) of an angle are the two lines that make it up.
  Interior The interior of an angle is the space between the sides of the angle, extending out to infinity.
  Exterior All the space on the plane that is not the interior.

Angles are measured in *degrees (°)*. For example, a whole round is equal to 360°.

Bisector of an angle

Given an angle, its *bisector* is the semi-straight line that, passing by its vertex, divides the angle into two identical angles.

Types of angles

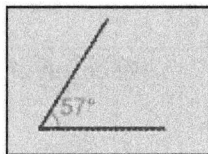

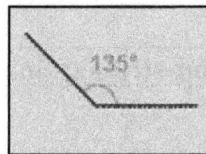

<u>Acute angle</u> Less than 90°	<u>Right angle</u> Exactly 90°	<u>Obtuse angle</u> Between 90° and 180°
<u>Straight angle</u> Exactly 180°	<u>Reflex angle</u> Between 180° and 360°	<u>Full angle</u> Exactly 360°

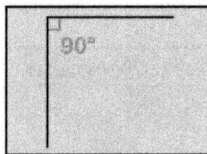

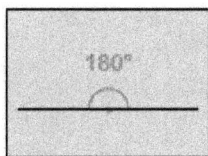

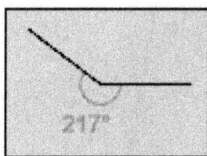

Angle relationships

Complementary angles: Two angles that add up to 90°.

Supplementary: Two angles that add up to 180°.

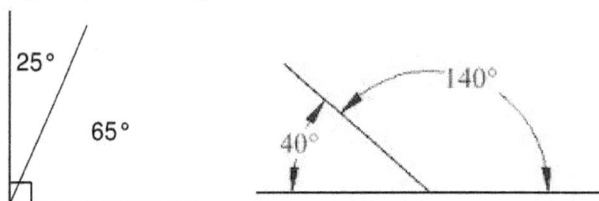

Complementary and supplementary angles

Consecutive angles: Two angles that share a common side and a common vertex, but do not overlap.

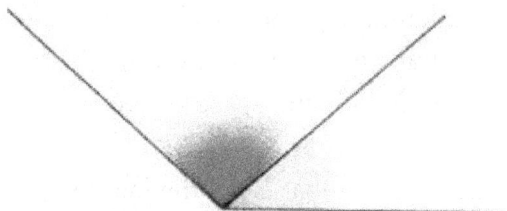

Adjacent angles: Two angles that share a common side and vertex, being the other two sides on the same straight line.

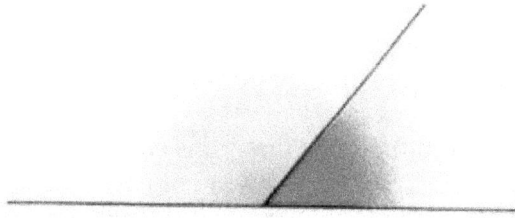

Vertical angles: A pair of <u>non-adjacent angles</u> formed by the intersection of two straight lines. Two vertical angles have the same measurement.

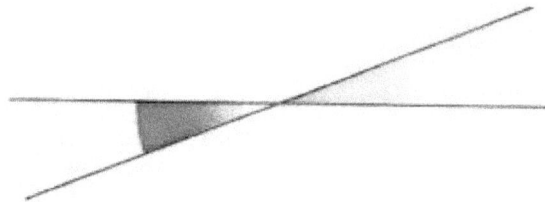

Vertical angles are **equal.**

Corresponding angles: Corresponding angles are created where a transversal line crosses a pair of parallel lines.

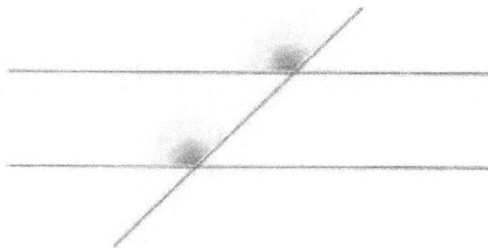

Corresponding angles are **equal.**

These angles are sometimes named **F** angles.

Alternate Angles: They can be classified into external and internal alternate angles.

External alternate

Alternate angles are **equal.**

These angles are sometimes named **Z** angles.

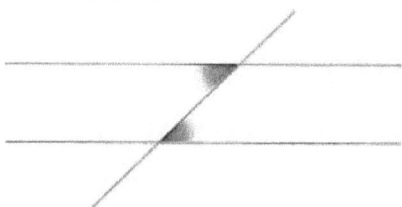

Internal alternate

3. Circle theorems

A straight line joining two points on the circumference is a **chord**.

An **arc** is the part of the circumference that joins two points.

We are going to study two important laws in circle angles.

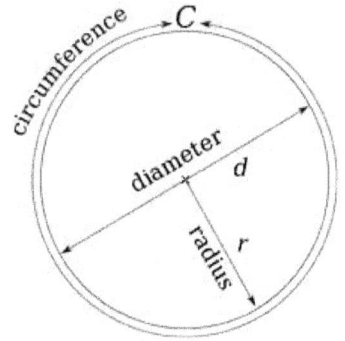

The **central angle** of a circle is twice the angle at the circumference from the same arc.	
Angles for the same arc in the same segment are equal.	

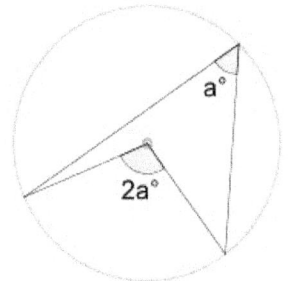

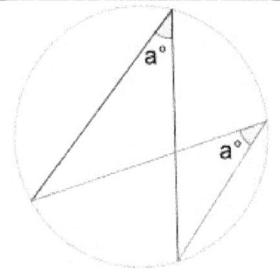

Example 1: Find the missing angles. Give a reason for each answer.

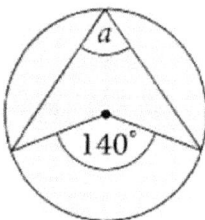

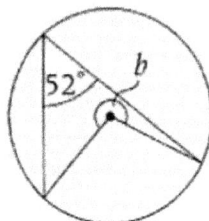

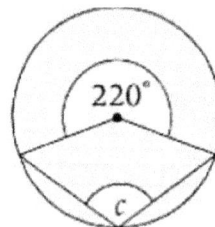

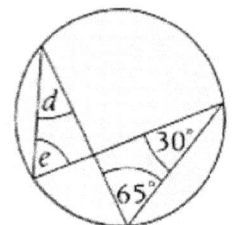

Solution:

a) a = 70° ↔ angle at the center is double the angle at the circumference.

b) b = 360° − 104° = 256° ↔ angles at a point.

c) c = 110° ↔ angle at the center is double the angle at the circumference.

d) d = 30° and e = 65° ↔ angles on same arc are equal.

176

Review exercises

Construction of geometric elements

1. Build an angle of 60º.

2. Build a triangle having three angles of 60º.

3. Build a triangle having angles of 30°, 60° and 90°.

4. Build a triangle having angles of 45°, 45° and 90°.

5. Build an angle having 120° and another one having 150°.

6. Draw a segment and build its bisector. What property do their points have?

7. Draw an angle and build its bisector.

8. Draw, with a protractor, angles of 30°, 45°, 60° and 75°. Build their complementary angles and calculate them.

9. Draw, with a protractor, angles of 120°, 135°, 150° y 165°. Build their supplementary angles and calculate them.

10. In your notebook, draw a straight line *r* and a point *P*, exterior to it. How many straight lines, parallel to *r* and passing by *P* could you draw? Build your drawings by using your technical instruments.

11. Answer the following questions:
   a) What property has each of the points of the bisector of a segment?
   b) In what point of the railway should we locate a train station so that it is as far from town A as from town B?

Angle relationships

12. Calculate the unknown angles in the following figures:

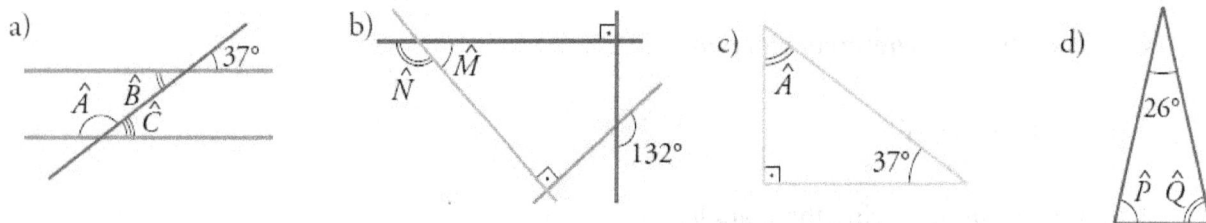

a)

b)

c)

d)

Sol.: a) $A = 143°$; $B = C = 37°$; **b)** $M = 48°$; $N = 132°$; **c)** $A = 53°$; **d)** $P = Q = 77°$.

13. Calculate the unknown angles in the following figures:

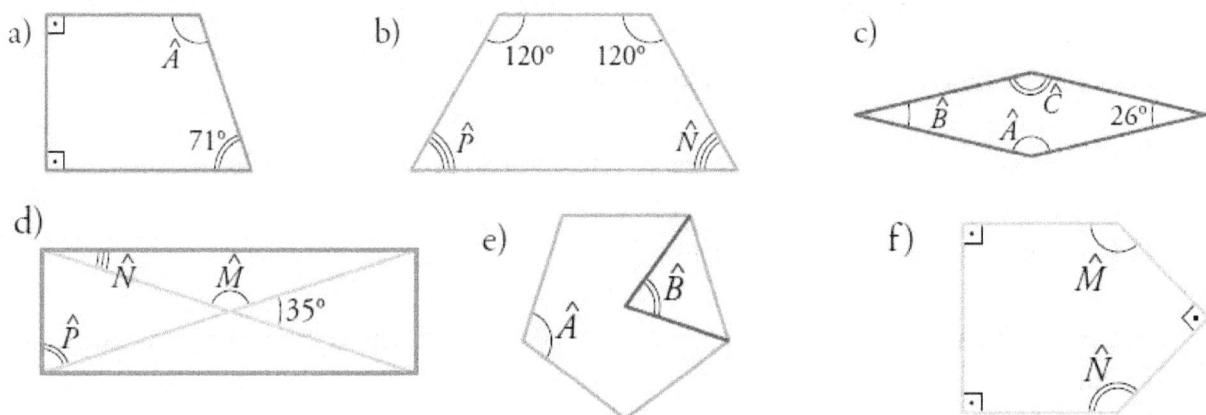

a)

b)

c)

d)

e)

f)

Sol.: a) $A = 109°$; **b)** $P = 60°$; **c)** $B = 26°$; $A = C = 154°$;
d) $N = 17° 30'$; $M = 145°$; $P = 72° 30'$; **e)** $A = 108°$; $B = 72°$; **f)** $M = N = 135°$.

14. Calculate indicated angles:

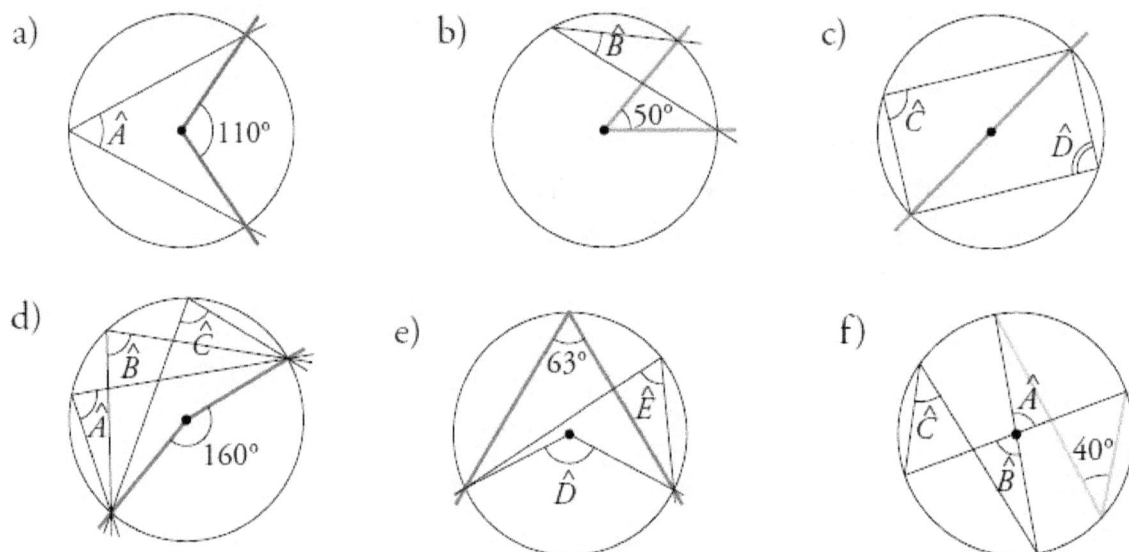

a)

b)

c)

d)

e)

f)

Sol.: a) $A = 55°$; **b)** $B = 25°$; **c)** $C = D = 90°$
d) $A = B = C = \dfrac{160}{2} = 80°$; **e)** $D = 126°$; $E = 63°$; **f)** $A = B = 80°$; $C = 40°$.

15. Calculate the red indicated angle of the following regular pentagon, by answering the following questions:

a) Central angle =
b) Red indicated angle =

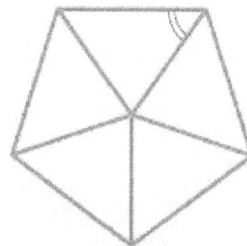

Sol.: a) 72°; b) 54°.

16. Calculate the unknown angle in each figure:

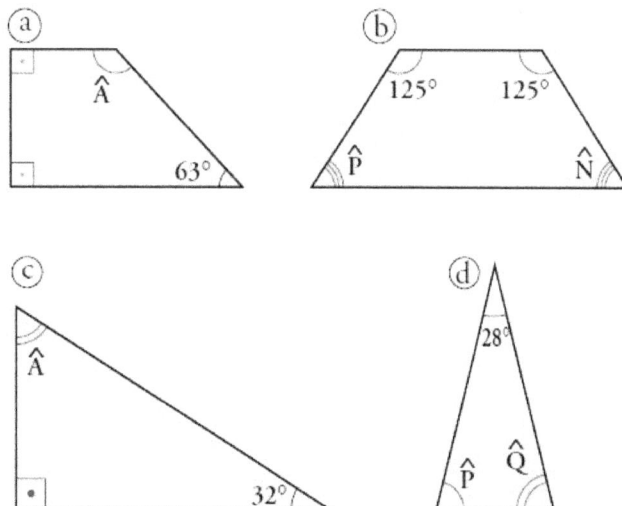

ⓐ $\hat{A}$ 63°

ⓑ 125°   125° $\hat{P}$ $\hat{N}$

ⓒ $\hat{A}$ 32°

ⓓ 28° $\hat{P}$ $\hat{Q}$

Sol.: a) A = 117°; b) P = N = 55°; c) A = 58°; d) P = Q = 76°.

17. Calculate the unknown angle in each figure:

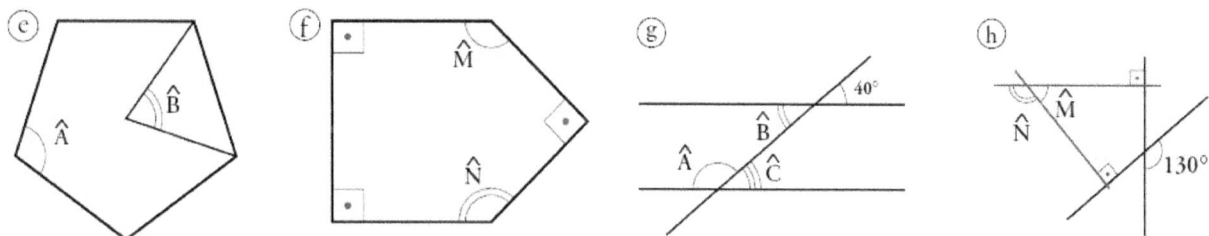

ⓔ $\hat{B}$ $\hat{A}$

ⓕ $\hat{M}$ $\hat{N}$

ⓖ 40° $\hat{B}$ $\hat{A}$ $\hat{C}$

ⓗ $\hat{N}$ $\hat{M}$ 130°

Sol.: e) A = 108; B =72°;   f) M = N = 135°; g) B = C = 40°; A = 140°; h) N = 130°; M = 50°.

179

Unit 9.- Plane shaped figures

1. Polygons

Polygons are closed figures formed by straight lines.

Polygons can be classified according to different criteria.

For example, polygons can be classified according to their number of sides:

TRIANGLE	QUADRILATERAL	PENTAGON	HEXAGON

Regular polygons

Regular polygons are those that have all their sides and angles identical.

Elements of a regular polygon:

- **Center:** Point from which all vertices are at the same distance.
- **Radius:** Segment joining center with a vertex.
- **Apothem:** Segment joining center with central point of a side.
- **Central angle:** angle formed by two consecutive radius.

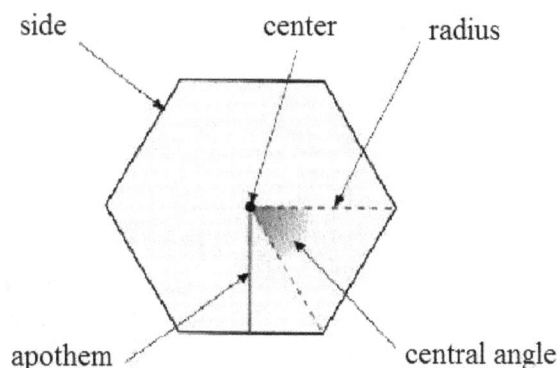

TRIANGLE	SQUARE	PENTAGON	HEXAGON
120°	90°	72°	60°

2. Triangles

Triangles, polygons having three sides, can be classified according to several criteria.

According to its sides, a triangle can be:

- **Equilateral:** If it has all its sides identical.
- **Isosceles:** If it has two of its sides identical, and another different one.
- **Scalene:** If it has its three sides different.

EQUILATERAL	ISOSCELES	SCALENE

According to its angles, a triangle can be:

- **Acute triangle:** If it has all its angles acute.
- **Right triangle:** If it has a right angle.
- **Obtuse triangle:** If it has an obtuse angle.

ACUTE	RIGHT	OBTUSE

Remarkable straight lines and angles of a triangle

Mediatrix	Bisector	Height	Median
Perpendicular straight line passing by the central point of a side.	Straight line passing by a vertex and divides the angle into two identical angles.	Perpendicular straight line to each side, passing by the opposite vertex.	Straight line passing by a vertex and the central point of the opposite side.
Circuncenter	**Incenter**	**Orthocenter**	**Barycenter**
Intersection point of mediatrices	Intersection point of bisectors	Intersection point of heights or their prolongations.	Intersection point of medians

182

Angles in a polygon

Central angle: It is the anlge formed by two consecutive radius.

Interior angles are inside the polygon.

Exterior angles are made by extending each side in the same direction.
Exterior angles are outside the polygon.

Sum of interior angles of a polygon

For a polygon with n sides the *interior angle sum = (n − 2) × 180°*

Example: Check previous rule for a) a triangle and b) a square.

<u>Solution:</u>

a) A triangle has n = 3 sides, so, its interior angles sum is $(3 - 2) \cdot 180 = 180°$, as you already knew.

b) You know interior angles sum for a square is 360°. Let's check it: n = 4 sides, so, $(4 - 2) \cdot 180 = 360°$.

Angles in a circumference

The **central angle** of a circle is twice the angle at the circumference from the same arc.	
Angles for the same arc in the same segment are equal.	

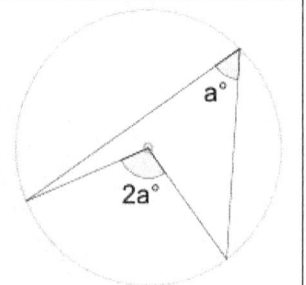

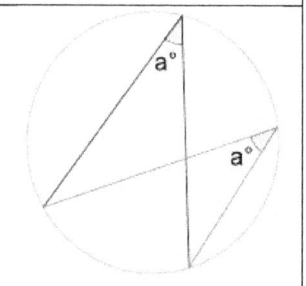

4. Quadrilaterals

A *quadrilateral* is a polygon having four sides.

Classification of quadrilaterals

PARALLELOGRAMS Quadrilaterals having all their sides parallel.	TRAPEZIA Quadrilaterals having two sides parallel.	TRAPEZOIDS Quadrilaterals that do not have any parallel side.

Classification of parallelogram quadrilaterals

FOUR IDENTICAL SIDES		IDENTICAL PARALLEL SIDES	
SQUARE (Right angles)	RHOMBUS (Opposite angles are identical)	RECTANGLE (Right angles)	RHOMBOID (Opposite angles are identical)

Classification of trapezium quadrilaterals

RIGHT TRAPEZIUM	ISOSCELES TRAPEZIUM

Review exercises

1. Classify the following polygons into regular and non-regular:

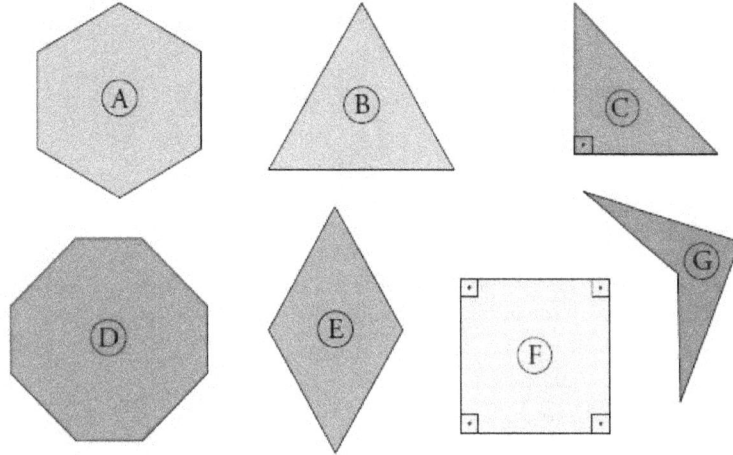

Sol.: Regular: A, B, F: Non-regular: C, D, E, G.

2. Classify the following triangles into acute, right and obtuse triangles:

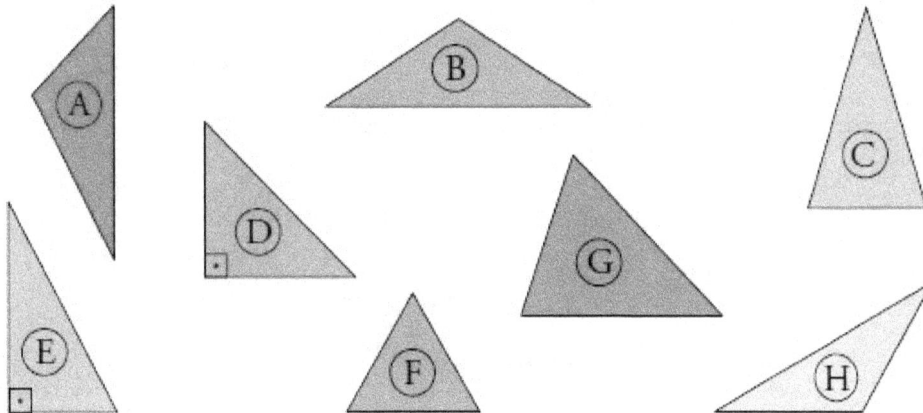

Sol.: Acute: C, F, G; Right: D, E; Obtuse: B, H.

3. Classify the following triangles according to their sides and their angles:

	a)	b)	c)	d)	e)	f)
SIDES						
ANGLES						

4. Classify the following triangles according to their sides and their angles:

Sol.: A: Obtuse isosceles; B: Acute isosceles; C: Acute equilateral; D: Right isosceles.

5. Name each quadrilateral:

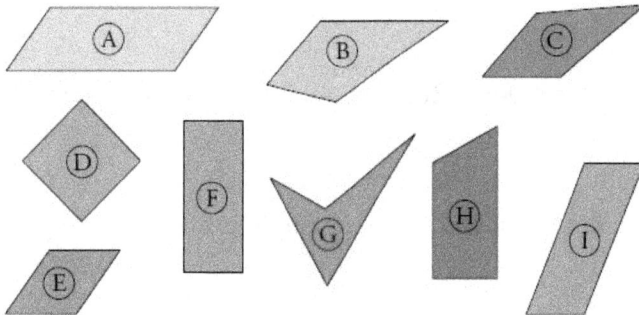

Sol.: A: Rhomboid, parallelogram. B: Trapezoid. C: Trapezoid. D: Square, parallelogram.
E: Rhombus, parallelogram. F: Rectangle, parallelogram. G: Trapezoid.
H: Right trapezium. I: Rhomboid, parallelogram.

6. In a triangle, two of its angles are A = 50° and 70°. Calculate the third angle.   **Sol.:** 60°.

7. Draw a triangle with sides a = 7 cm, b = 5 cm and c = 8 cm. With a protractor, measure its angles and check their sum is correct.   **Sol.:** 60°, 82° and 38°.

8. Try to draw a triangle with sides 10 cm, 5 cm and 3 cm. What conclusion do you extract?

9. Build a triangle with sides a = 6 cm, b = 3 cm, and angle between them, C = 110°. Measure the other elements.   **Sol.:** 7.6 cm, 42° and 28°.

10. Build a triangle with sides a = 8 cm, B = 70°, C = 50°. Measure the other elements.
                Sol.: 7.1 cm, 8.6 cm and 80°.

11. Draw a triangle with sides a = 6 cm, b = 8 cm and c = 12 cm. Draw its medians and locate its barycenter. Check that barycenter divides each median into two segments, being one twice the other one.

186

12. Draw a triangle with sides a = 8 cm, b = 10 cm and c = 12 cm. Draw its heights and locate its orthocenter.

13. Draw a triangle with sides a = 10 cm, b = 8 cm and c = 6 cm. Observe it is a right triangle. Locate its orthocenter and measure the height on the hypotenuse, h_a. Check that $a \cdot h_a = b \cdot c$.

14. Draw a triangle with sides a = 12 cm, b = 8 cm and c = 6 cm. Observe it is an obtuse triangle. Draw its heights, prolongate them and observe they intersect in an exterior point.

15. Where is the orthocenter of a right triange, inside or outside it?

Sol.: Inside, because it is acute.

16. Draw a triangle with sides 3 cm, 4 cm and 5 cm. Locate its circumcenter and draw the circunscribed circumference. Observe the triangle is a right one and that circumcenter is on the hypotenuse.

17. Draw a triangle with sides 3 cm, 5 cm and 7 cm. Locate its circumcenter and draw the circunscribed circumference. Observe the triangle is an obtuse one and that circumcenter is outside it.

18. Draw a triangle with sides 5 cm, 7 cm and 8 cm. Locate its incenter and draw the inscribed circumference.

19. If in a right triangle, an angle is 20° 25', what is the other acute angle?   **Sol.:** 69° 35'.

21. If in a rhombus, an angle is 20°, What are the other angles?   **Sol.:** 20°, 160° and 160°.

22. Calculate the other angles:

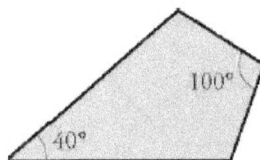

Sol.: 20° and 20°.

23. Calculate the unknown angles in the following figures:

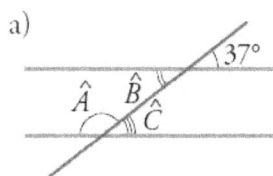

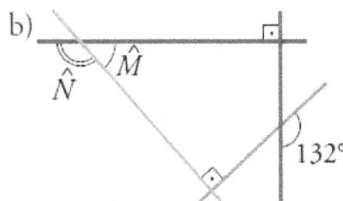

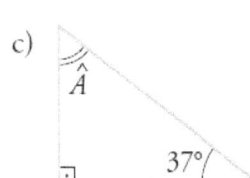

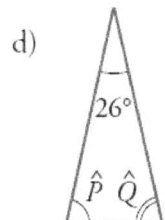

Sol.: a) A = 143°; $B = C$ = 37°;   b) M = 48°; N = 132°;   c) A = 53°;   d) $P = Q$ = 77°.

187

24. Calculate the unknown angles in the following figures:

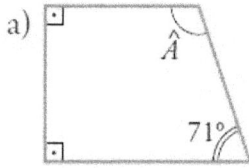

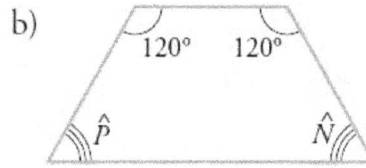

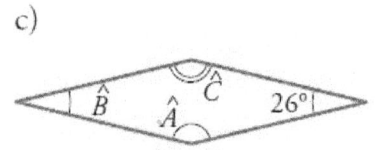

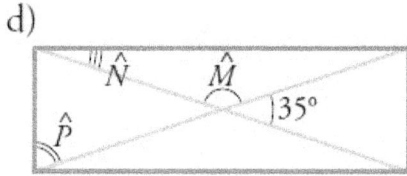

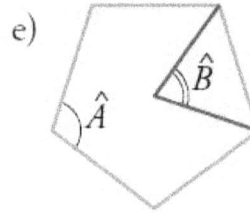

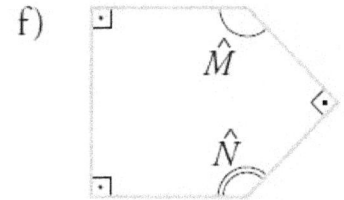

a)

b)

120°   120°

$\hat{P}$ $\hat{N}$

c)

$\hat{B}$ $\hat{A}$ $\hat{C}$ 26°

$\hat{A}$

71°

d)

$\hat{N}$ $\hat{M}$

$\hat{P}$ 35°

e)

$\hat{B}$

$\hat{A}$

f)

$\hat{M}$

$\hat{N}$

Sol.: a) $A = 109°$; b) $P = 60°$; c) $B = 26°$; $A = C = 154°$;
d) $N = 17° 30'$; $M = 145°$; $P = 72° 30'$; e) $A = 108°$; $B = 72°$; f) $M = N = 135°$.

25. Calculate the red indicated angle of the following regular pentagon, by answering the following questions:

a) Central angle =
b) Red indicated angle =

Sol.: a) 72°; b) 54°.

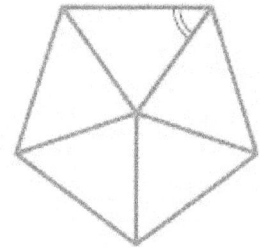

26. Calculate the unknown angle in each figure:

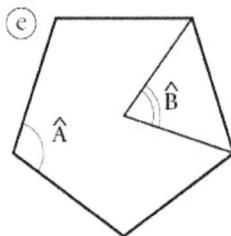

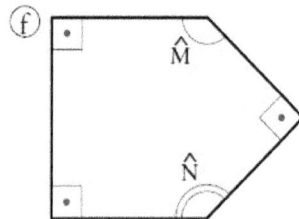

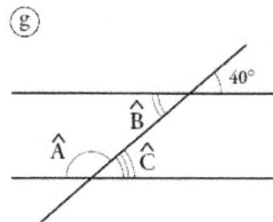

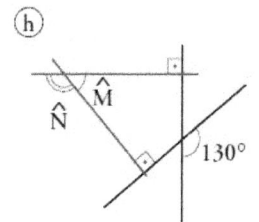

e)

$\hat{B}$

$\hat{A}$

f)

$\hat{M}$

$\hat{N}$

g)

40°

$\hat{B}$

$\hat{A}$ $\hat{C}$

h)

$\hat{M}$

$\hat{N}$

130°

Sol.: e) A = 108; B =72°; f) M = N = 135°; g) B = C = 40°; A = 140°; h) N = 130°; M = 50°.

188

27. Calculate the indicated angles:

a)

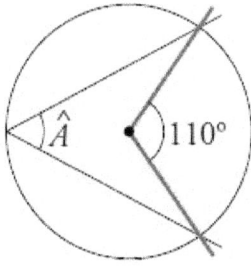

b)

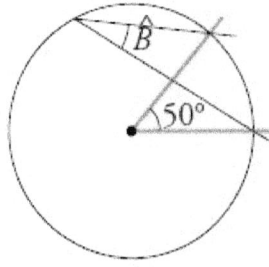

c)

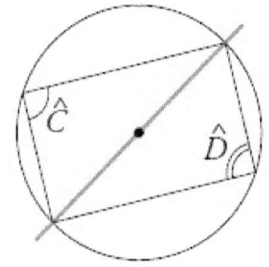

d)

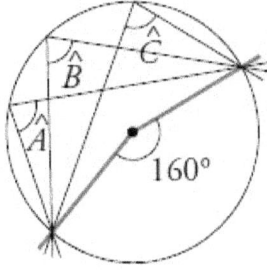

e)

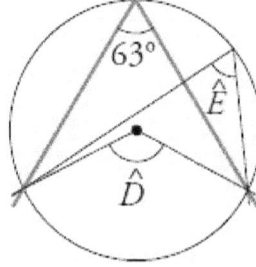

f)

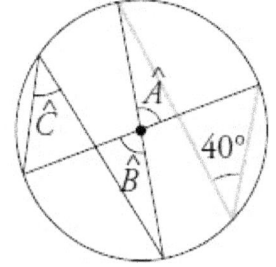

Sol.: a) $A = 55°$; b) $B = 25°$; c) $C = D = 90°$

d) $A = B = C = \dfrac{160}{2} = 80°$; e) $D = 126°$; $E = 63°$; f) $A = B = 80°$; $C = 40°$.

28. Calculate the unknown angle in each figure:

(a)

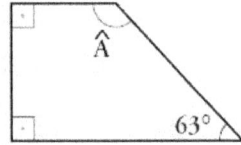

(b)

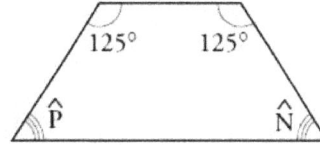

(c)

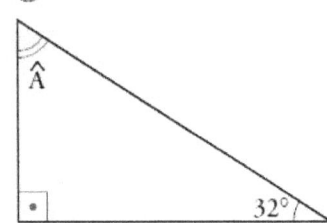

(d)

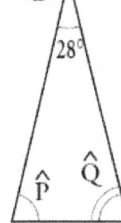

Sol.: a) A = 117°; b) P = N = 55°; c) A = 58°; d) P = Q = 76°.

Unit 10.- Geometry

1. Similarity

Two figures that have the same shape are said to be **similar**. There are lots of examples of similar figures in quotidian life: an island and its map, a person and his photograph, etc.

> Two shapes are said to be *similar shapes* if:
> - corresponding pairs of angles are equal
> - corresponding pairs of sides are in the same *ratio*.

You can use the ratio between similar shapes to find missing side lengths.

Example: Find unknown sides of following figures, knowing they are similar.

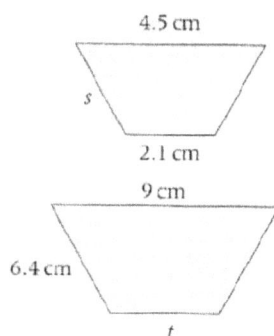

Solution:

The ratio larger/smaller is $\dfrac{9}{4.5} = 2$

Side s corresponds to side 6.4 cm $\rightarrow$ $\dfrac{6.4}{s} = 2 \Rightarrow s = \dfrac{6.4}{2} = 3.2\,cm$

Side t corresponds to side 2.1 cm $\rightarrow$ $\dfrac{t}{2.1} = 2 \Rightarrow t = 2.1 \cdot 2 = 4.2\,cm$

1. These two parallelograms are similar. Find the perimeter of the smaller parallelogram.

12 cm

6 cm

5 cm

2. Measures of a photograph are 15 cm x 10 cm and it is surrounded by a frame being 2 cm wide. Are interior and exterior rectangles similar?

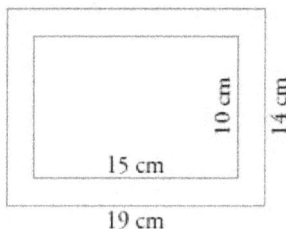

10 cm

14 cm

15 cm

19 cm

Scales. A ratio is used in scale drawings of maps and buildings.

Scale of a drawing = Drawing length : Actual length

A scale is usually expressed without mentioning units as in 1 : 100 000

A scale of 1 : 100 000 means that the real distance is 100 000 times the length of 1 unit on the map or drawing.

3. A particular map shows a scale of 1 : 500. What is the actual distance if the map distance is 8 cm?

4. A particular map shows a scale of 1 cm : 5 km. What would the map distance (in cm) be if the actual distance is 14 km?

5. Actual distance of 900 km is represented on a map by 6 cm. What is its scale?

2. Similar triangles

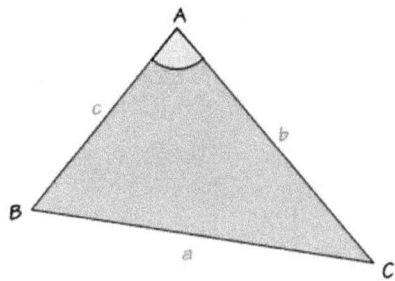

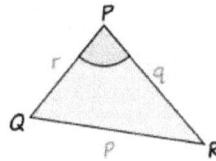

In similar triangles:

- corresponding pairs of angles are equal

$$A = P , \ B = Q , \ C = R$$

- corresponding pairs of sides are in the same ratio

$$\frac{a}{p} = \frac{c}{r} = \frac{b}{q}$$

If a line is drawn parallel to one side of a triangle, the smaller triangle and the larger one are *similar*. These two triangles are said to be in a ***Thales position***.

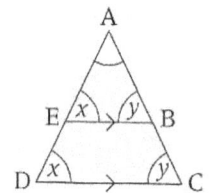

Example: Find the length CD in this triangle.

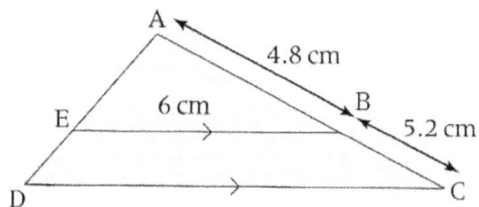

Solution:

Triangle ACD is similar to triangle ABE

$$\frac{CD}{6} = \frac{10}{4.8} \Rightarrow CD = \frac{6 \cdot 10}{4.8} = 12.5 \ cm$$

Exercises

6. EB is parallel to DC. Work out the perimeter of: a) the triangle ABE and b) the trapezium EBCD.

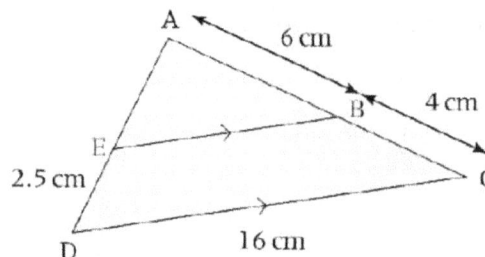

193

How to decide if two triangles are similar?

Two triangles are similar if they have all their corresponding pairs of angles equal and corresponding sides are in the same ratio. But we don't have to know all the sides and all the angles … two or three out of the six is enough.

There are three ways to find if two triangles are similar:

> - If two triangles have two of their angles equal, the triangles are similar.
> - If two triangles have two of their sides in the same ratio and the included angles are also equal, then the triangles are similar.
> - If two triangles have three pairs of sides in the same ratio, then the triangles are similar.

7. Check that the following pairs of triangles are similar.

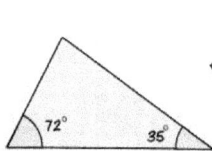

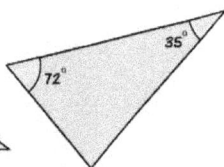

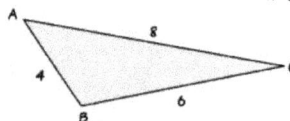

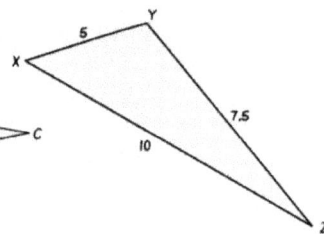

3. Pythagoras' Theorem

As you know, a **right triangle** or a *right angled triangle* is a triangle with one right (90°) angle.

The **hypotenuse** of a right triangle is the side that is opposite the right angle.

The **legs** of a right triangle are the sides adjacent to the right angle.

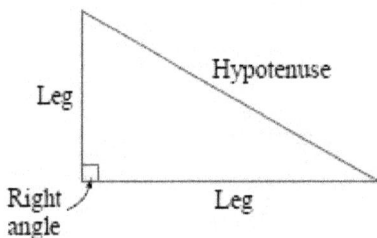

When drawing a right triangle, right angle is indicated by a little square (□), in order to avoid doubts about its measure.

Thousands of years ago, a Greek named **Pythagoras** discovered that the lengths of the sides of a right triangle have a special relationship. This relationship is known as the Pythagorean Theorem.

The *Pythagorean Theorem* enables us to calculate the length of the third side of a right triangle when we know the lengths of the other two sides.

Pythagorean Theorem

In a right triangle, the sum of the squares of the legs is equal to the square of the hypotenuse

$$(\text{leg})^2 + (\text{leg})^2 = (\text{hypotenuse})^2 \qquad \text{or} \qquad a^2 + b^2 = c^2$$

Example: To hold a 1.5 m stick, we use a wire to ankle it to the floor. Look at the figure and calculate the length of the wire we need.

Solution: $\quad \text{leg}_1 = 1.5 \quad \text{leg}_2 = 2.6 \quad \text{hyp} = x$

$$(1.5)^2 + (2.6)^2 = x^2 \rightarrow 2.25 + 6.76 = x^2$$

$$x^2 = 9.01 \quad \rightarrow \quad x = \sqrt{9.01} \approx \boxed{3 \text{ m}}.$$

Example: Look at the figure and calculate the height at which it is flying.

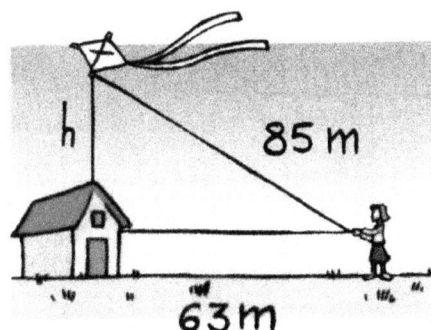

Solution: $\quad \text{leg}_1 = 63 \quad \text{leg}_2 = h \quad \text{hyp} = 85$

$$(63)^2 + x^2 = 85^2 \rightarrow 3969 + x^2 = 7225$$

$$x^2 = 7225 - 3969 = 3256 \quad \rightarrow \quad x = \sqrt{3256} \approx \boxed{57 \text{ m}}.$$

8. Calculate the length of unknown sides:

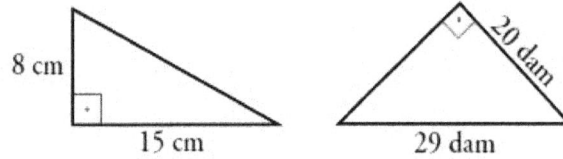

8 cm

15 cm

20 dam

29 dam

9. Diagonals of a rhombus measure 16 cm and 12 cm. Calculate its side.

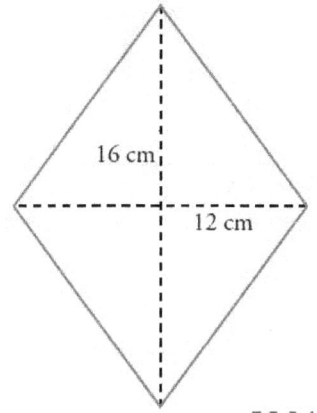

16 cm

12 cm

10. In following right triangles, one leg and hypotenuse are given and we want to know the other leg.

    a) 45 cm and 37 cm   b) 39 cm and 15 cm.

11. In a rhombus, we know one diagonal, 24 cm, and its side, 13 cm. Find out the other diagonal.

12. Lower sides of a triangle are 5 and 12 cm. Calculate the third one if we want it to be a right triangle. What type of triangle will it be if third side is 10 cm?

How to know if a triangle is a right one

Pythagoras' Theorem can be used to determine if a triangle is acute, right or obtuse:

> If a and b are the shortest sides and c, the longest one:
> - If $a^2 + b^2 < c^2$, it is an **acute** triangle.
> - If $a^2 + b^2 = c^2$, it is a **right** triangle.
> - If $a^2 + b^2 > c^2$, it is an **obtuse** triangle.

13. Classify into right angled, acute or obtuse, the triangles with sides:

a) 7 cm, 8 cm, 11 cm

b) 11 cm, 17 cm, 15 cm

c) 34 m, 16 m, 30 m

d) 65 m, 72 m, 97 m.

14. Find out the radius of this circumference (Fig. 1), knowing: OP = 39 cm, PT = 36 cm.

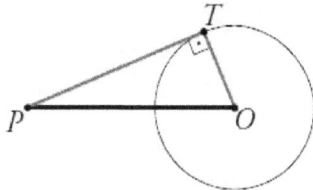

Figure 1

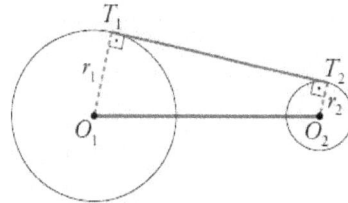

Figure 2

15. In Figure 2, $r_1 = 15$ cm, $r_2 = 6$ cm, $O_1O_2 = 41$ cm. Calculate the length of T_1T_2.

16. Check if triangle with sides 29 cm, 35 cm and 48 cm is right, acute or obtuse. Calculate the length of its height on the largest side.

17. Calculate the unknown side.

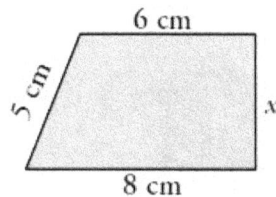

18. A trapezium has bases of 26 cm and 36 cm, and lateral sides measure 13 cm. calculate its height.

19. Knowing the height of the building, a = 108 m, the distance from P to its base, d 45 m, calculate the length of the wire.

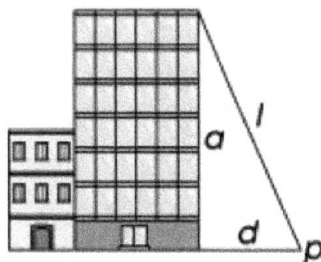

Area of a polygon is quantity of surface existing into the closed zone delimited by sides. Some of the expressions used to calculate areas of polygons might be known by you. They are shown in the next figure. *YOU DO NOT NEED TO MEMORIZE THEM!!!*

Anyway, we will calculate areas of polygons (regular or irregular) by cutting them into three basic shapes: rectangles, triangles and circles (or portions of them, as semicircles, etc).

KNOWN AREAS

RECTANGLE	SQUARE	PARALLELOGRAM	RHOMBUS
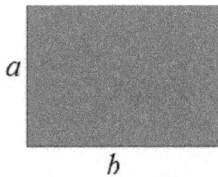		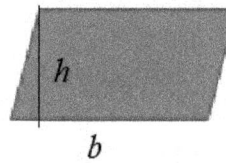	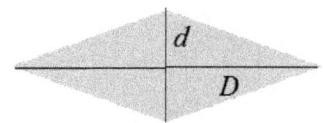
$A = b \cdot a$	$A = l^2$	$A = b \cdot a$	$A = \dfrac{D \cdot d}{2}$

TRAPEZIUM	REGULAR POLYGON	TRIANGLE	RIGHT TRIANGLE
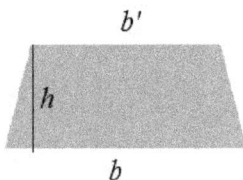		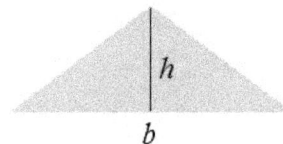	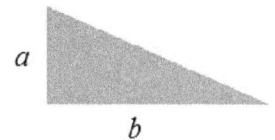
$A = \dfrac{b + b'}{2} \cdot h$	$A = \dfrac{Per \cdot ap}{2}$	$A = \dfrac{b \cdot h}{2}$	$A = \dfrac{b \cdot a}{2}$

Example: Calculate the perimeter, the length of the diagonal and the area of a room whose dimensions are 8.3 m x 4.6 m.

Solution: In geometrical problems, it is very useful to draw a scheme:

- Perimeter: P = 2 · 8.3 + 2 · 4.6 = $\boxed{25.8 \text{ m}}$.

- Area: A = 8.3 · 4.6 = $\boxed{38.18 \text{ m}^2}$.

- To calculate the diagonal, we use Pythagoras' theorem:

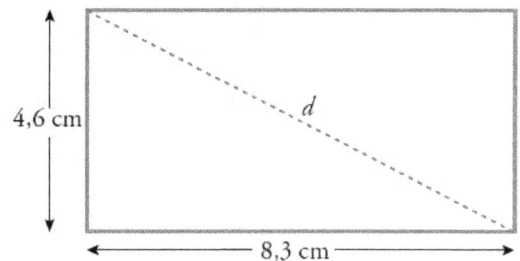

$$(8.3)^2 + (4.6)^2 = x^2 \rightarrow x^2 = 90.05 \rightarrow \qquad x = \sqrt{90.05} \approx \boxed{9.5 \text{ m}}.$$

20. Calculate the perimeter, the length of the diagonal and the area of square with side 15 cm.

21. Calculate the height of a rectangle with area 40 m^2 and base 5 m.

22. Calculate the perimeter of a rectangle with area 60 m^2 and height 12 m.

23. Calculate the area of the following figures:

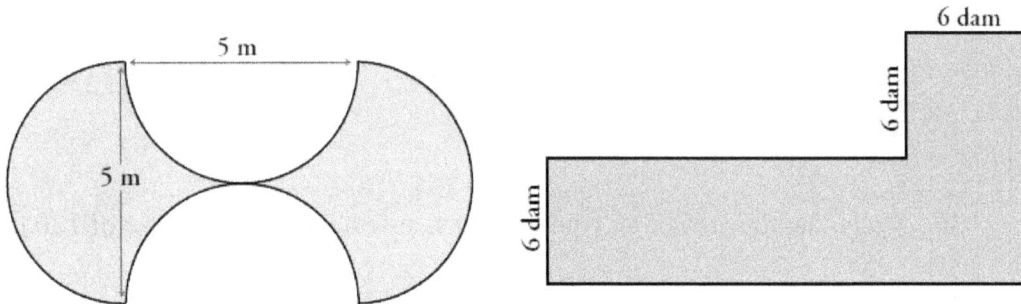

24. Calculate the perimeter and area of these parallelograms:

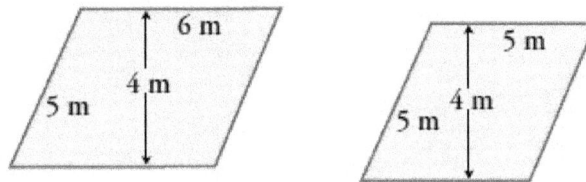

25. Calculate the area of this parallelogram and the value of x:

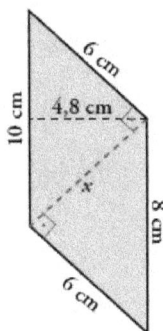

26. Calculate the area and the perimeter of a rhombus, knowing its diagonals are 10 cm and 24 cm

27. Calculate the area of a rhombus with perimeter is 40 cm and its bigger diagonal is 16 cm

28. Calculate the area of this triangular garden.

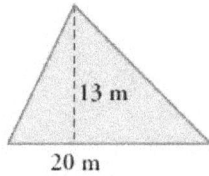

13 m

20 m

29. Calculate the area of this isosceles triangle.

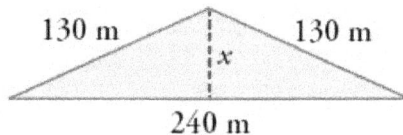

130 m   130 m

x

240 m

30. Calculate the area of an equilateral triangle having a perimeter of 120 m.

31. A right triangle has legs of 18 and 24 cm. Calculate:

a) Its area   b) Its perimeter   c) The height on the hypotenuse.

32. Calculate the area of this trapezium.

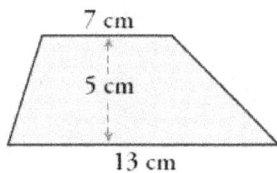

7 cm

5 cm

13 cm

33. Calculate the area of a trapezium whose bases measure 12 and 20 m and has a height of 10 m.

34. Calculate the area and the perimeter of a right trapezium with bases of 16 cm and 11 cm and inclined side of 13 cm.

35. Calculate the area and perimeter of an isosceles trapezium having bases of 20 cm and 36 cm, and a height of 15 cm

36. Calculate the area of the following figure:

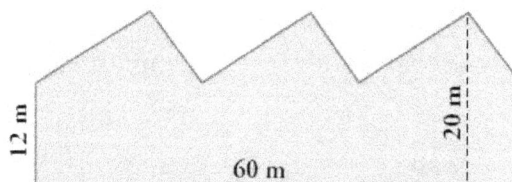

12 m

20 m

60 m

200

5. Areas of curved figures

KNOWN AREAS

| CIRCLE | CIRCLE SECTOR | ANNULUS (Circle crown) | ELLIPSE |

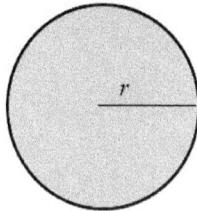
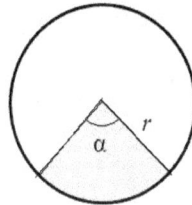
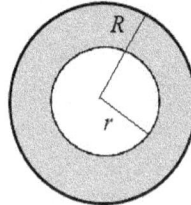
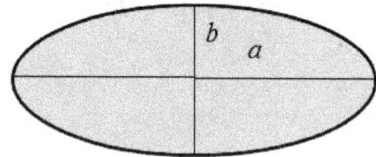

$$A = \pi \cdot r^2$$

$$A = \pi \cdot r^2 \cdot \frac{\alpha}{360}$$

$$A = \pi \cdot (R^2 - r^2)$$

$$A = \pi \cdot a \cdot b$$

Exercises

37. Calculate the area and perimeter of this coloured figure:

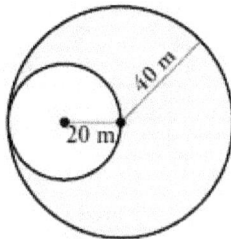

38. Calculate the perimeter and area of this coloured figure:

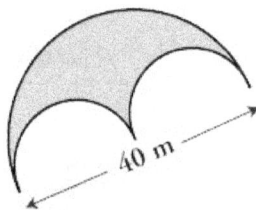

39. Calculate the perimeter and area of this coloured figure:

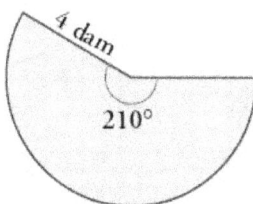

40. Calculate the length of an arc of circumference with radius of 10cm and angle of 40°.

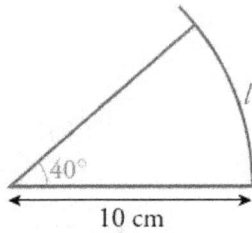

41. Calculate the area and perimeter of this figure:

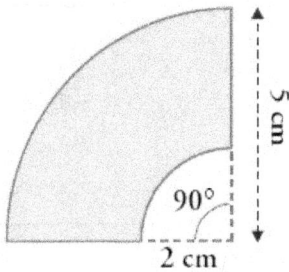

6. Areas of solids

In this section, you will calculate the area of the polygons forming the faces of solids.

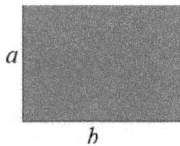

$A = b \cdot a$

$A = l^2$

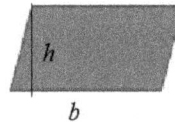

$A = b \cdot a$

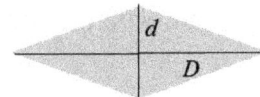

$A = \dfrac{D \cdot d}{2}$

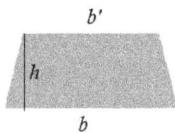

$A = \dfrac{b + b'}{2} \cdot h$

$A = \dfrac{Per \cdot ap}{2}$

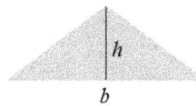

$A = \dfrac{b \cdot h}{2}$

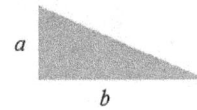

$A = \dfrac{b \cdot a}{2}$

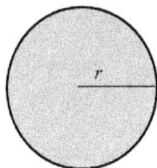

$A = \pi \cdot r^2$

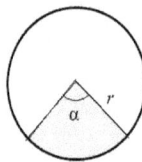

$A = \pi \cdot r^2 \cdot \dfrac{\alpha}{360}$

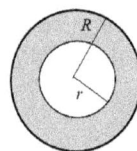

$A = \pi \cdot (R^2 - r^2)$

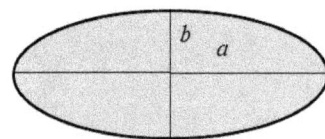

$A = \pi \cdot a \cdot b$

202

And two important solids: Sphere and Cone:

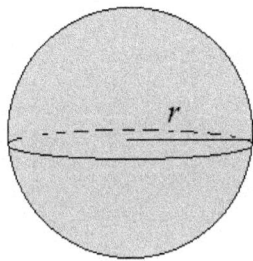

$$A = 4 \cdot \pi \cdot r^2$$

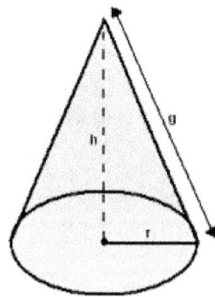

$$A = \pi \cdot r \cdot (g + r)$$

42. Calculate the area of these polyhedrons, obtained from cubes of edge 12 cm.

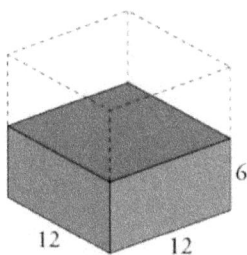

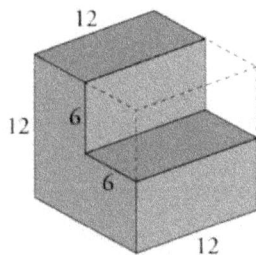

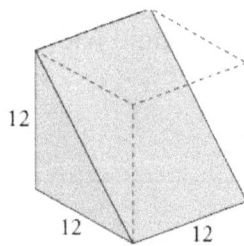

43. Calculate the area of the prism and the pyramid. In both cases, base is a regular hexagon.

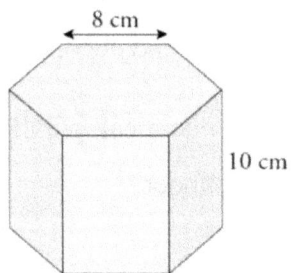

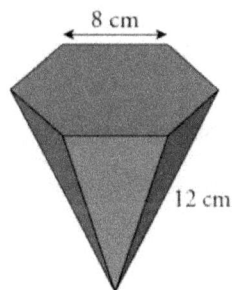

44. Calculate the area of these solids:

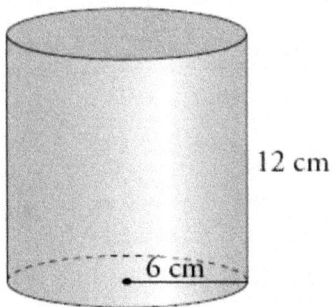

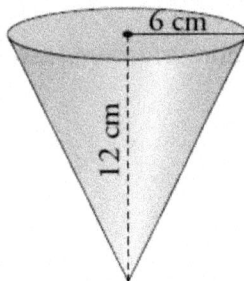

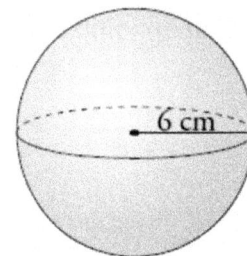

45. Calculate the area of these solids.

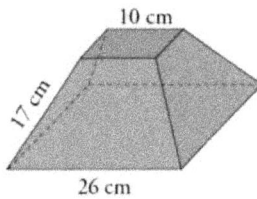

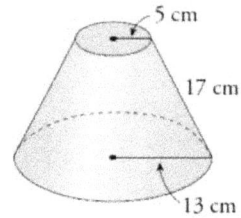

46. Calculate the areas of:

   a) A right prism whose base is a rhombus of diagonals 12 cm and 20 cm, knowing that

   its lateral edge measures 24 cm.

   b) A right pyramid with the same base and edge that previous prism.

7. Volume of solids

In order to organize correctly this part of the unit and make it easier to you, we are going to divide solids into three groups: right, appointed and curved solids.

- Having two bases: Those having two identical parallel bases. Their representatives are prism and cylinder.

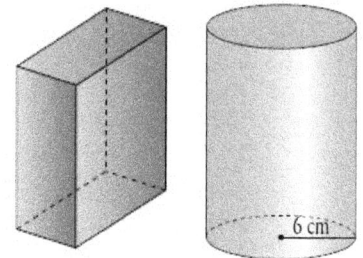

- Having 1 base and 1 upper vertex: Those having a base and an appointed head. Their representatives are pyramid and cone.

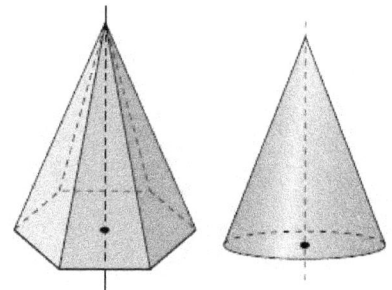

- Curved solids: Sphere and those derived from sphere.

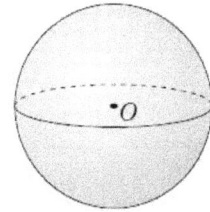

Corresponding solids: As you can see, a pyramid can be imagined into a prism, or a cone into a cylinder. We are going to name these pairs of solids, *corresponding solids*.

Now, study of solids' volume is going to be very systematical:

SOLID	VOLUME
2 bases	$V = A_{base} \cdot h$
1 base, 1 vertex	$V = \dfrac{V_{CORRESPONDING}}{3}$ or $V = \dfrac{A_{base} \cdot h}{3}$
Sphere	$V = \dfrac{4}{3} \cdot \pi \cdot R^3$

47. Calculate the volume of these solids, obtained from prisms.

48. Calculate the volume of these solids, whose bases are regular polygons.

49. Calculate the volume of these solids.

Review exercises

Similarity

1. On a map whose scale is 1:1 500 000, distance between two towns is 3,5 cm.

a) What is the actual distance between them?

b) What is the distance, on that map, between two towns at 250 km?

2. Calculate actual distance that is represented by 13 cm on a map with scale 1: 1800000.

3. Calculate the lengths of the sides *a* and *b* knowing these triangles have their sides parallel:

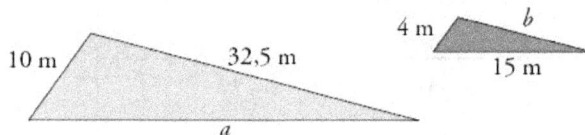

4. Thales was able to determine the height of a pyramid only by measuring its shadow, 18 m, and the shadow projected by his stick, 0.5 m, at the same time. What was its height if his stick was 1.70 m?

Pythagoras' theorem

5. Classify into right angled, acute or obtuse, the triangles with sides:

a) a = 61 m, b = 60 m, c = 11 m   b) a = 18 cm, b = 15 cm, c = 12 cm

c) a = 30 m, b = 24 m, c = 11 m   **Sol.:** a) Right; b) Acute; c) Obtuse.

6. Calculate the height at which this globe is located, knowing the wire measures 50 m.

Sol.: 40 m.

7. Calculate the perimeter of triangle ABC, rounding the result to the tenths.

Sol.: 10.5 cm.

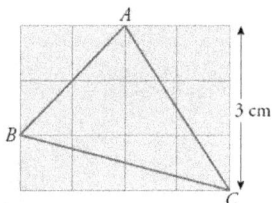

8. Calculate the unknown side of these right triangles.   **Sol.:** 39 m and 65 dm.

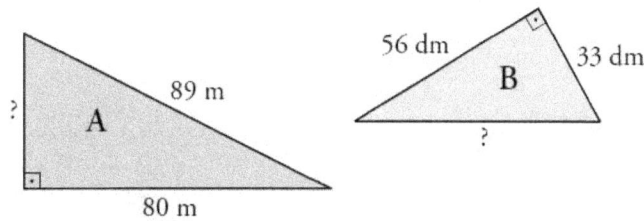

9. Calculate the unknown sides of these right triangles, rounding the result to the tenths:

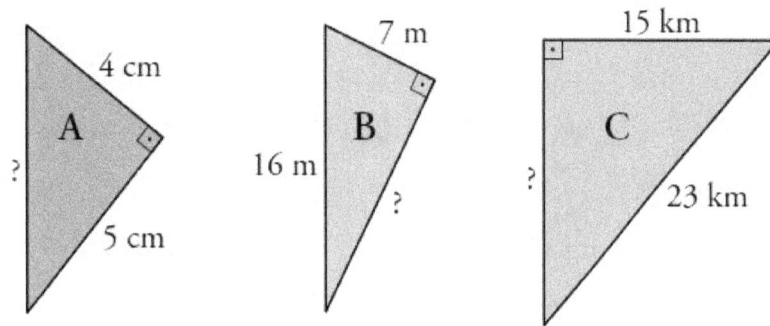

Sol.: 6.4 cm, 14.4 m and 17.4 km.

10. What is the side of the square whose diagonal measures 6 cm?   **Sol.:** 4.2 cm.

11. Diagonal of a rectangle measures 20 cm and one of its sides, 12 cm. What is the other side?

Sol.: 16 cm.

12. Calculate the side of a rhombus whose diagonals measure 6 and 8 cm.   **Sol.:** 5 cm.

13. In a rhombus, a diagonal measures 16 cm and its side 17 cm. What is the other diagonal?

Sol.: 30 cm.

14. In a right trapezium, parallel sides measure 17 and 11 cm, and its inclined side, 10 cm. What is its height?   **Sol.:** 8 cm.

15. Remember in a hexagon, lengths of side and radius are equal. Calculate the apothem of an hexagon having side 6 cm.   **Sol.:** 5.2 cm.

16. Side of a regular pentagon measures 12 cm and its radius, 10.2 cm. Calculate its apothem with one decimal digit.   **Sol.:** 8.2 cm.

17. Radius of a regular pentagon is 20 cm and its apothem, 16.2 cm. Calculate its side with one decimal digit.   **Sol.:** 23.4 cm.

18. Side of a regular octagon measures 8 cm and its apothem 9.6 cm. Calculate the radius of its circumscribed circumference.   **Sol.:** 10.4 cm.

19. Calculate, with one decimal digit, the height of an equilateral triangle with side 12 cm.

Sol.: 10.4 cm.

20. A straight line passes at 18 cm of the centre of a circumference of radius 19.5 cm. Does it intersect the circumference? What is the length of the chord that they determine?   **Sol.:** 15 cm.

Areas and perimeters

21. Calculate the area and perimeter of the following figures:

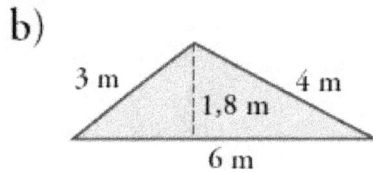

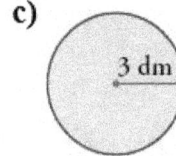

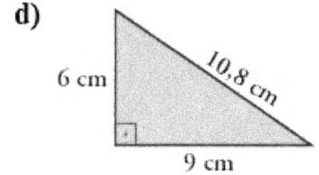

Sol.: a) 9 m^2; 12 m;   b) 5.4 m^2; 13 m; c) 28.26 dm^2; 18.84 dm; d) 27 cm^2; 25.8 cm.

22. Calculate the area and perimeter of the following figures:

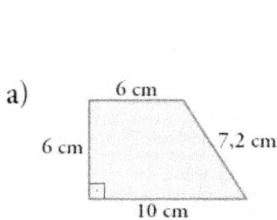

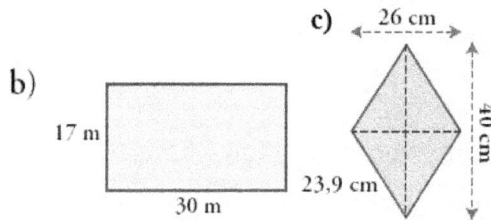

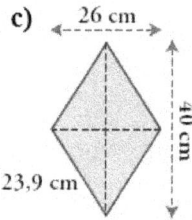

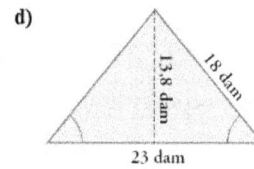

Sol.: a) 48 cm^2; 29.2 cm;   b) 510 m^2; 94 m; c) 520 cm^2; 95.6 cm; d) 158.7 dam^2; 59 dam.

23. Calculate the area and perimeter of the following figures:

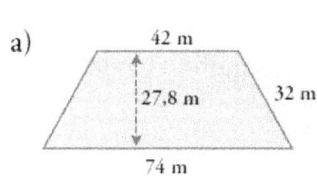

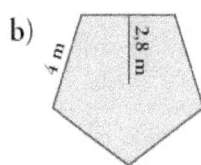

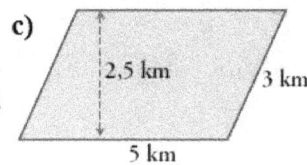

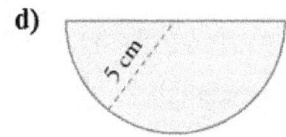

Sol.: a) 1612.4 m^2; 180 m;   b) 28 m^2; 20 m; c) 12.5 km^2; 16 km; d) 39.25 cm^2; 25.7 cm.

24. Calculate the area and perimeter of the following figures:

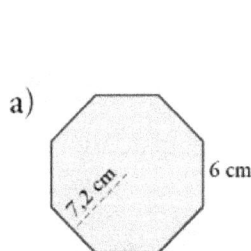

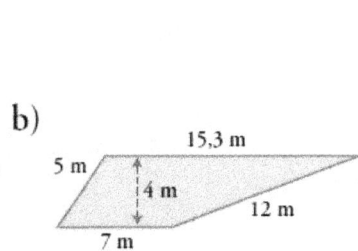

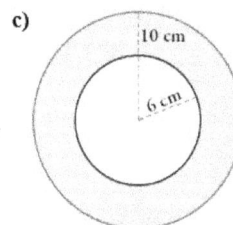

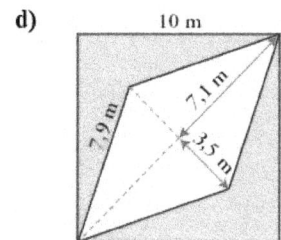

Sol.: a) 172.8 cm^2; 48 cm;   b) 44.6 m^2; 39.3 m; c) 200.96 cm^2; 100.48 cm; d) 50.3 m^2; 71.6 m.

208

25. Calculate the area and perimeter of the following figures:

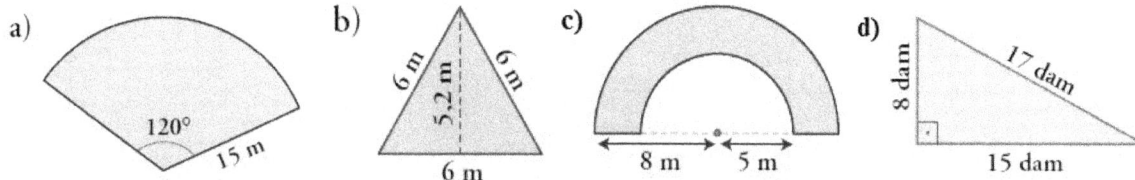

a)

b)

c)

d)

120°

15 m

6 m

5,2 m

6 m

6 m

8 m   5 m

8 dam

17 dam

15 dam

Sol.: a) 235.5 m²; 61.4 m;   b) 15.6 m²; 18 m; c) 61.23 m²; 46.82 m; d) 60 dam²; 40 dam.

26. Calculate the unknown element, the area and the perimeter of the following figures:

a)

b)

4 m

3 m

Sol.: a) 120°, 16.7 m² and 16.4 dam;   b) 4.2 m, 27.1 m² and 45.2 m.

27. Calculate the area and the perimeter of the following figure:

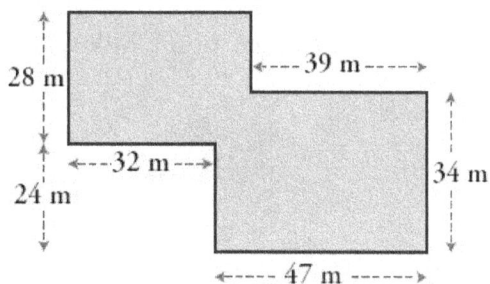

28 m

39 m

32 m

24 m

34 m

47 m

Sol.: 2638 m² and 262 m.

28. Calculate the area and the perimeter of the following figures:

a)

b)   2 cm

8 cm

5 cm

13 cm

14 cm

Sol.: a) 24 cm² and 20 cm; b) 40 cm² and 34 cm.

29. Calculate the area and the perimeter of the following figure:

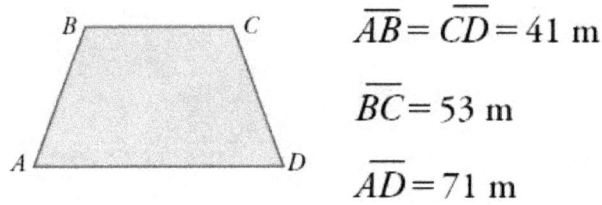

$$\overline{AB} = \overline{CD} = 41 \text{ m}$$

$$\overline{BC} = 53 \text{ m}$$

$$\overline{AD} = 71 \text{ m}$$

Sol.: 2480 m^2 and 206 m.

30. Calculate the area and the perimeter of the following figure:

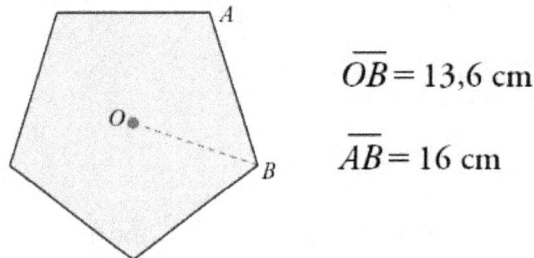

$$\overline{OB} = 13,6 \text{ cm}$$

$$\overline{AB} = 16 \text{ cm}$$

Sol.: 440 cm^2 and 80 cm.

31. Calculate the area and the perimeter of the following figure:

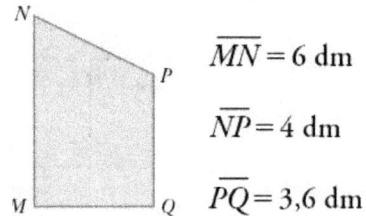

$$\overline{MN} = 6 \text{ dm}$$

$$\overline{NP} = 4 \text{ dm}$$

$$\overline{PQ} = 3,6 \text{ dm}$$

Sol.: 15.4 dm^2 and 16.8 dm.

32. Calculate the area and the perimeter of the following figure:

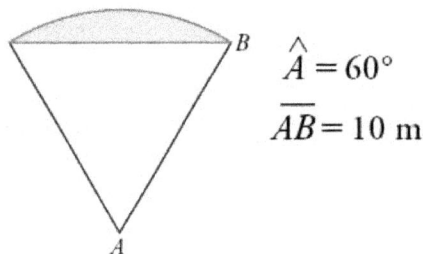

$$\hat{A} = 60°$$

$$\overline{AB} = 10 \text{ m}$$

Sol.: 8.8 m^2 and 20.5 m.

33. Calculate the area and the perimeter of the following figure:

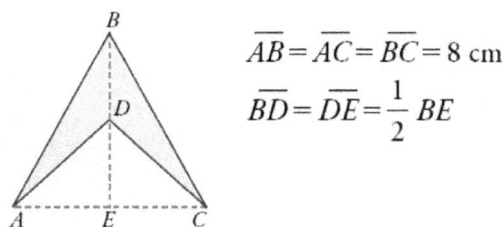

$$\overline{AB}=\overline{AC}=\overline{BC}=8 \text{ cm}$$
$$\overline{BD}=\overline{DE}=\frac{1}{2}BE$$

Sol.: 13.8 cm^2 and 26.6 cm.

34. Calculate the area of the following figure:

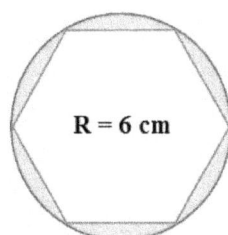

R = 6 cm

Sol.: 19.44 cm^2.

35. To cover a rectangular floor, we have used 175 stones of 20 dm^2 each. How many square shaped stones of side 50 cm will we need to cover a neighbor's identical floor? **Sol.:** 140 stones.

36. A rhombus has an area of 24 cm^2 and one of its diagonal is 8 cm. Calculate its perimeter.

Sol.: 20 cm.

37. knowing that the side of the square is 30 cm, calculate the radius of the inscribed and circumscribed circumferences and the area of the shaded zone.

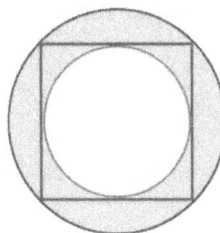

Sol.: 15 cm, 21.2 cm and 704.7 cm^2.

38. Is this octagon regular? Calculate its area and its perimeter.

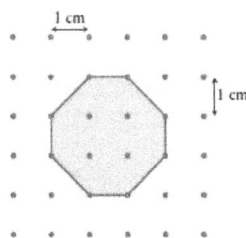

1 cm

1 cm

Sol.: Not, 7 cm^2 ad 9.66 cm.

39. Calculated the shaded zone of these figures:

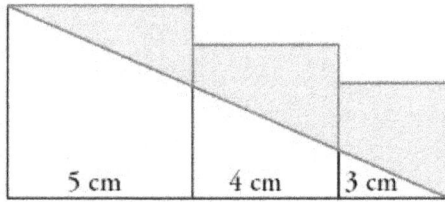 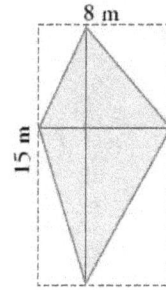

Sol.: 20 cm^2 and 60 m^2.

Areas and volumes of solids

40. Calculate the volume existing between the cube and the cone:

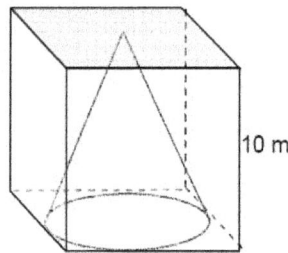

41. Calculate the area of this figure:

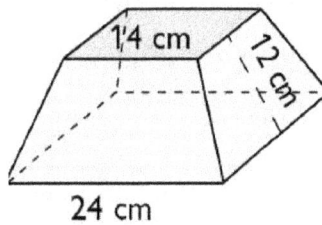

42. Calculate area and volume of first figure, and volume of second one:

a)

b)

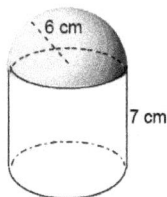 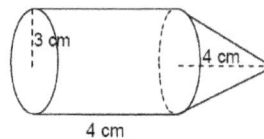

43. a) Estimate the volume and the area of the Earth, knowing its radius is, approximately, 6371 km.

b) Atmosphere has a mean height of 100 km. Estimate its volume and the percent of the total volume that it represents.

44. Calculate volume of Kio Towers, in Madrid, knowing its base is a square of side 35 m and its height is 114 m.

45. We want to paint walls and ceiling of a dining-room whose floor is 12 x 7 m, and whose height is 3.5 m. This room has two doors with dimensions 1 x 2 m, and three windows with dimensions 2 x 2 m.

   a) How much area is there to paint? (Draw it).

   b) If we have painting cans for 25 m^2, how many cans do we need?

46. Calculate the volume of a Rubik's cube with edge 8 cm.

47. Calculate the volume, in ml, of a can of Coca-cola, whose height is 10.9 cm and whose diameter is 6.2 cm. (1 ml = 1 cm^3)

48. Calculate the volume of Keops' pyramid, knowing its height is 230.35 m and its base is a square of side 136.86 m.

49. A water deposit is a prism whose height is 10 m and whose volume is 4000 m^3. Calculate the side of the base, knowing it is a square.

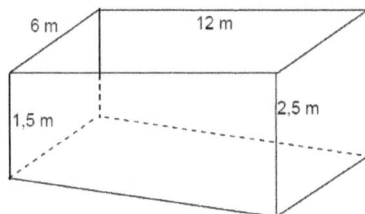

50. Calculate the volume of the shown swimming pool. (1 m^3 = 1 000 l)

51. About previous exercise, how long would it take to fill it if water comes at 0.5 l/s?

(EXERCISE 50)

52. Base diameter of a cylinder is equal to its height. Total area is 169.56 m^2. Calculate its dimensions.

53. Calculate the area of the following solids:

a)

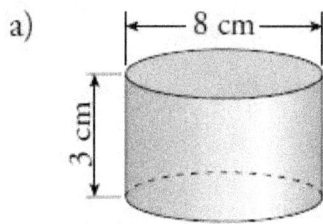

8 cm

3 cm

b)

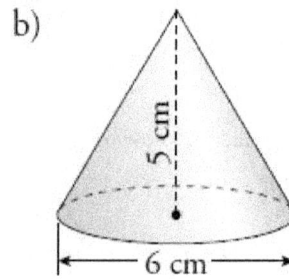

5 cm

6 cm

c)

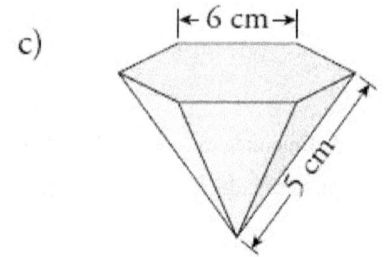

6 cm

5 cm

d)

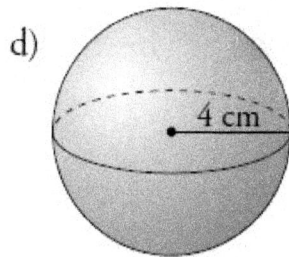

4 cm

e)

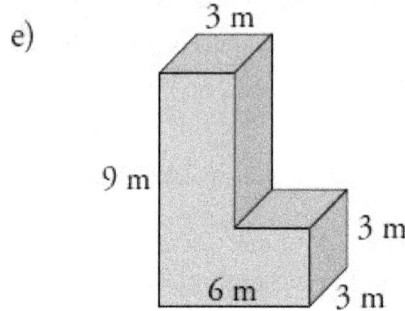

3 m

9 m

6 m

3 m

3 m

f)

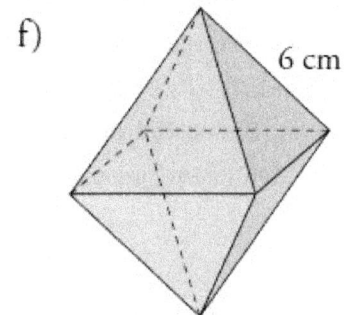

6 cm

54. Calculate the volume of the following solids:

a)

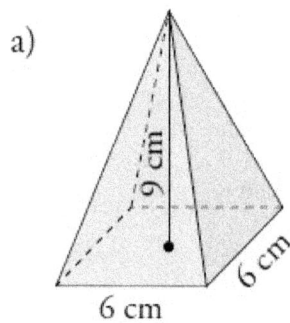

9 cm

6 cm

6 cm

b)

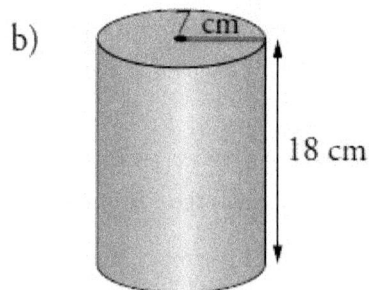

7 cm

18 cm

c)

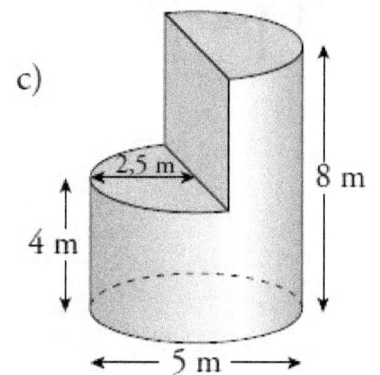

2,5 m

4 m

8 m

5 m

d)

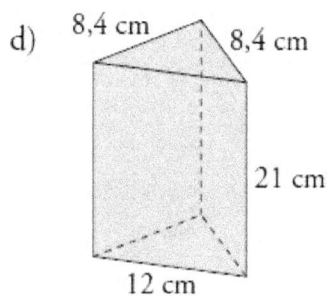

8,4 cm

8,4 cm

21 cm

12 cm

e)

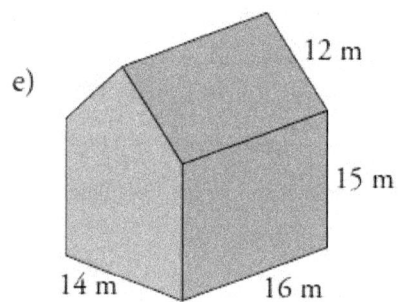

12 m

15 m

14 m

16 m

f)

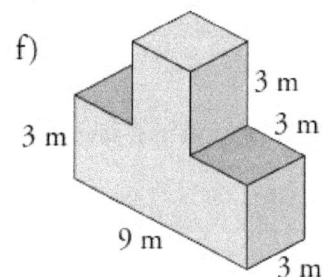

3 m

3 m

3 m

3 m

9 m

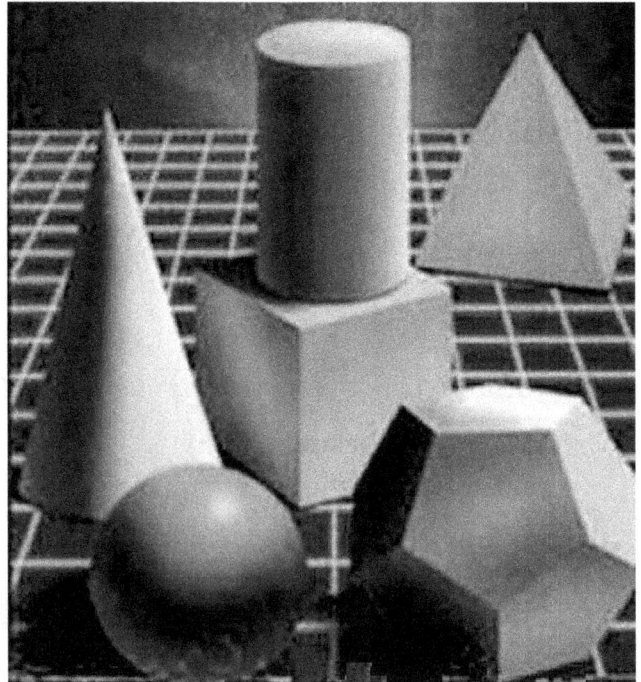

Unit 11.- Solids

1. Polyhedrons

In geometry, the word *solid* refers to the shell of a three-dimensional figure, not including its interior. Solids can be classified in many ways, based on their shape.

- **Polyhedron:** It is the solid delimited by polygons.

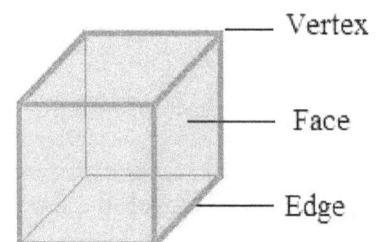

The parts of a polyhedron are:

- **Face:** Each polygon that forms the polyhedron.

- **Edge:** Intersection of two faces.

- **Vertex:** Intersection of two or more edges.

- **Vertex order:** Number of faces intersecting in a vertex.

Polyhedrons can be classified under different criteria:

    a) Faces being identical regular polygons or not.

    b) Dihedral angles.

    c) Number and shapes of their faces.

a) Regular and irregular polyhedrons: *Regular* polyhedron is that one in which all its faces are identical regular polygons and all its vertices have the same order. The rest of the polyhedrons are named *irregular*.

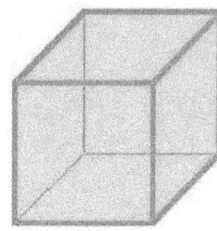

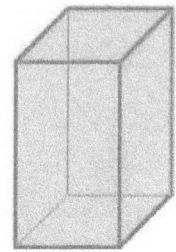

Regular   Irregular

There are only five convex *regular polyhedrons*, and they are known collectively as the **Platonic solids**, shown below.

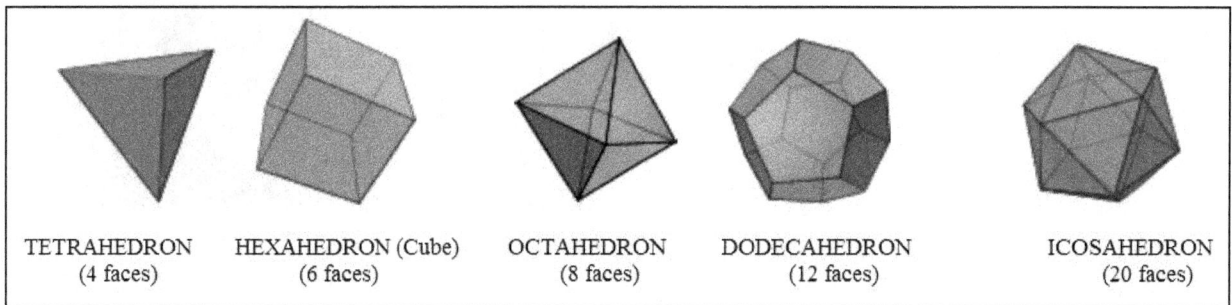

| TETRAHEDRON (4 faces) | HEXAHEDRON (Cube) (6 faces) | OCTAHEDRON (8 faces) | DODECAHEDRON (12 faces) | ICOSAHEDRON (20 faces) |

b) Convex and concave polyhedrons: A *convex* polyhedron is that in which all the dihedral angles are less than 180°. On the other hand, a *concave* polyhedron is that one that has, at least, one dihedral angle bigger than 180°.

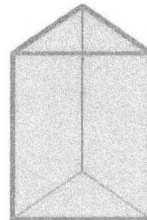

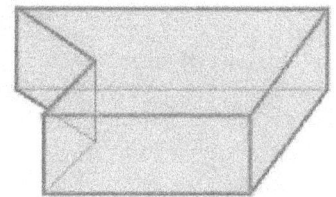

Convex   Concave

Exercises	**1.** Develop a 2 cm side octahedron.
	2. Develop a 4 cm side cube.
	3. Draw an oblique prism with a triangular base.

1.1. Prisms

A prism is a polyhedral with two identical parallel bases, whose lateral faces are parallelograms.

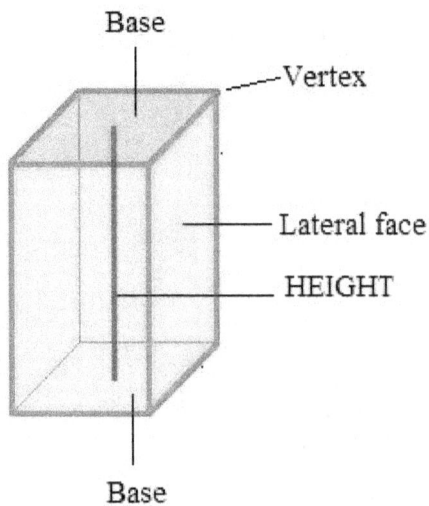

- The **height** is the perpendicular distance between the bases.
- Prisms are classified depending of the polygon in their bases, into *triangular*, *quadrangular*, etc.
- **Right**: All its lateral faces are rectangles (they are also perpendicular to the bases).
- **Oblique:** Not all its faces are rectangles (they are not perpendicular to the bases).

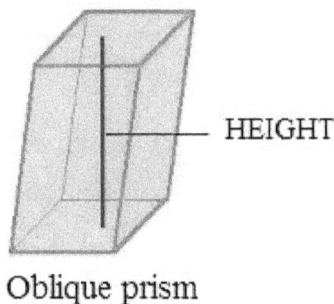

Oblique prism

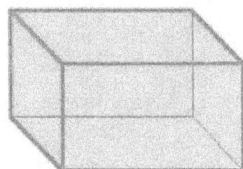

Parallelepiped

Orthohedron

- **Parallelepiped**: Prism whose bases are parallelograms.

- **Orthohedron**: Parallelepiped in which 6 faces re rectangles.

- **Regular Prisms**: Those that are right and their bases are regular polygons.

1.2. Pyramids

A *pyramid* is the polyhedron whose base is a polygon and its lateral faces are triangles converging in the vertex of the pyramid.

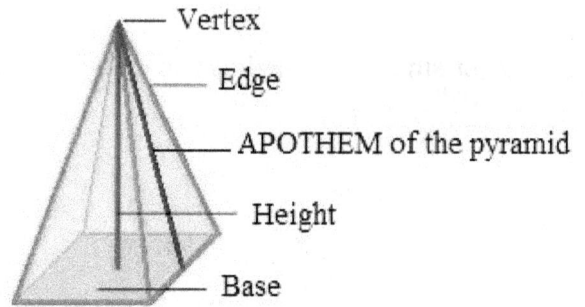

- **Height:** distance from the vertex to the base.

- **Right pyramid:** A pyramid whose lateral faces are isosceles triangles (Also: the height begins in the center off the base).

- **Oblique pyramid:** It is nor a right one.

- **Regular pyramid:** a right pyramid having in the base a regular polygon.

Oblique pyramid

- **Apothem of a regular pyramid:** height one of its faces.

Pyramids can be classified depending on the type of polygon having in their base, as *triangular*, *quadrangular*, etc.

Exercises	**4.** What is the polygon in the base of a pyramid if it has 5 lateral faces? **5.** What is the polygon in the base of a pyramid that has 8 faces? **6.** Draw the development of a pyramid having all its face identical. **7.** Which of the following figures is the development of a pyramid?

1.3. Right Cylinder

A *right cylinder* is the solid generated by rotating a rectangle about one of its sides.

- **Rotation axis:** side about which rectangle has rotated.

- **Generatrix:** side of the rectangle that is opposite the rotation axis.

- **Height:** (perpendicular) distance between the bases. Its bases are circles.

- **Radius** of the base is one of the perpendicular sides to the rotation axis.

- **Oblique cylinder:** it is the cylinder resulting of cutting a right cylinder by two parallel planes that are not perpendicular to the rotation axis.

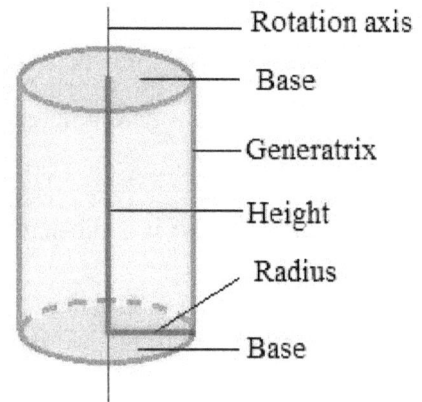

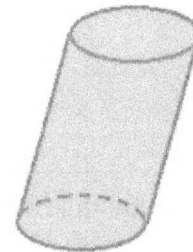

Oblique cylinder

1.4. Right cone

A *right cone* is the solid that is generated by rotating a right triangle about one of its legs.

- It has one **base** and one **vertex**.

- **Rotation axis:** leg about which right triangle rotates.

- **Generatrix:** hypotenuse of the right triangle.

- **Height:** Segment from the vertex to the base. It is the rotation axis.

- **Radius** of the base is the leg of the right triangle that is¡s not the rotation axis.

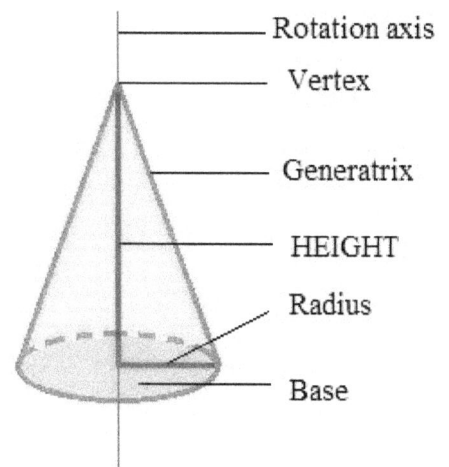

2.3. Sphere

A *sphere* is the solid that is generated by rotating a semicircle about its diameter.

- **Rotation axis:** it is the semicircle diameter.

- **Generatrix:** it is the semi-circumference of the semicircle.

- The **radius** of the sphere is the semicircle radius.

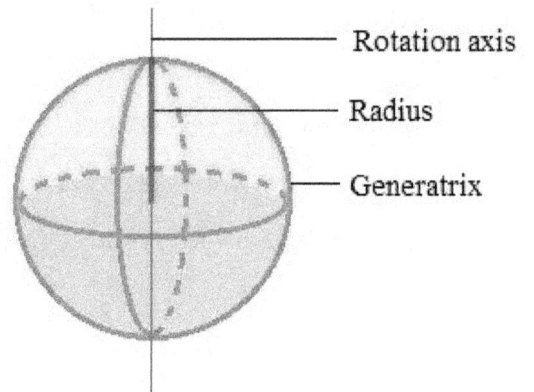

Rotation axis
Radius
Generatrix

Exercises

8. Name the following solids:

a)   b)   c)   d)   e)

f)   g)   h)   i)   j)

2. Areas of solids

Just to remember:

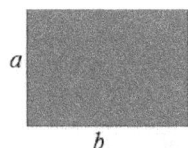

$$A = b \cdot a$$

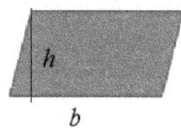

$$A = l^2$$

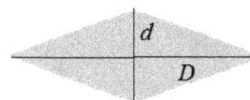

$$A = b \cdot a$$

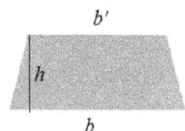

$$A = \frac{D \cdot d}{2}$$

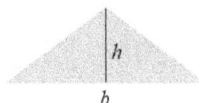

$$A = \frac{b + b'}{2} \cdot h$$

$$A = \frac{Per \cdot ap}{2}$$

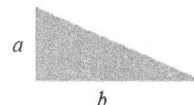

$$A = \frac{b \cdot h}{2}$$

$$A = \frac{b \cdot a}{2}$$

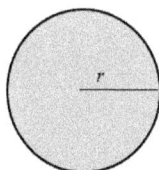

$$A = \pi \cdot r^2$$

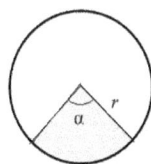

$$A = \pi \cdot r^2 \cdot \frac{\alpha}{360}$$

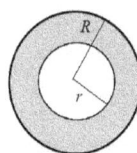

$$A = \pi \cdot (R^2 - r^2)$$

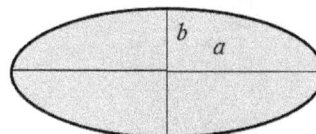

$$A = \pi \cdot a \cdot b$$

And two important solids: Sphere and Cone:

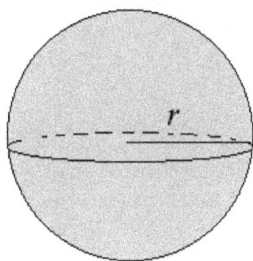

$$A = 4 \cdot \pi \cdot r^2$$

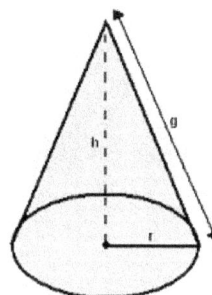

$$A = \pi \cdot r \cdot (g + r)$$

9. Calculate the area of these polyhedrons, obtained from cubes of edge 12 cm.

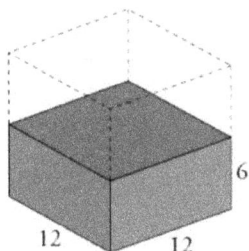

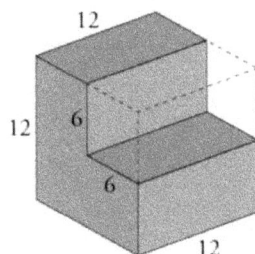

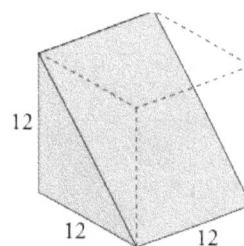

Exercises

221

10. Calculate the area of the prism and the pyramid. In both cases, base is a regular hexagon.

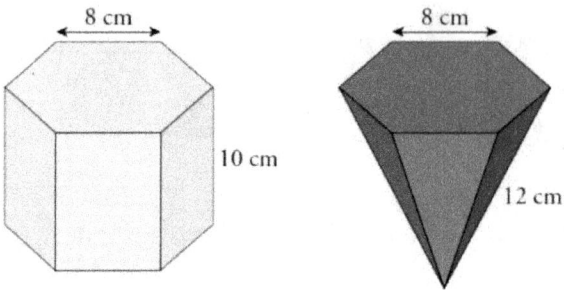

11. Calculate the area of these solids:

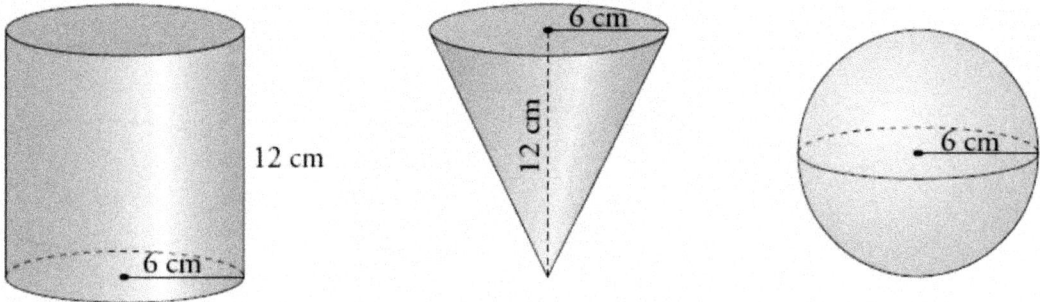

12. Calculate the area of these solids.

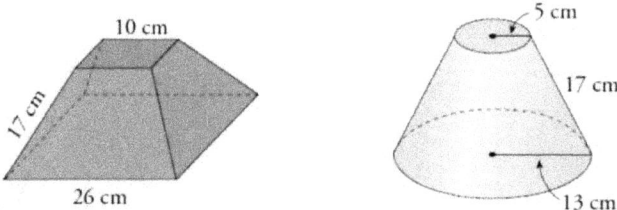

13. Calculate the areas of:

a) A right prism whose base is a rhombus of diagonals 12 cm and 20 cm, knowing that its lateral edge measures 24 cm.

b) A right pyramid with the same base and edge that previous prism.

3. Volume of solids

In order to organize correctly this part of the unit and make it easier to you, we are going to divide solids into three groups: right, appointed and curved solids.

- **Having two bases:** Those having two identical parallel bases. Their representatives are prism and cylinder.

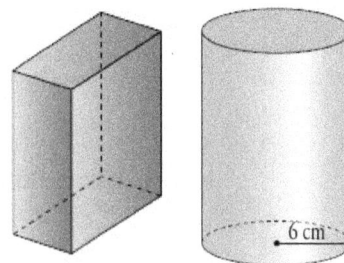

- **Having 1 base and 1 upper vertex:** Those having a base and an appointed head. Their representatives are pyramid and cone.

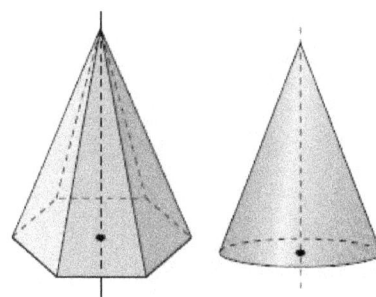

- **Curved solids:** Sphere and those derived from sphere.

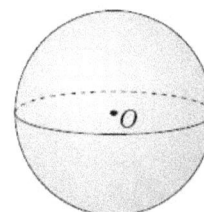

Corresponding solids: As you can see, a pyramid can be imagined into a prism, or a cone into a cylinder. We are going to name these pairs of solids, *corresponding solids*.

Now, study of solids' volume is going to be very systematical:

SOLID	VOLUME
2 bases	$V = A_{base} \cdot h$
1 base, 1 vertex	$V = \dfrac{V_{CORRESPONDING}}{3}$ or $V = \dfrac{A_{base} \cdot h}{3}$
Sphere	$V = \dfrac{4}{3} \cdot \pi \cdot R^3$

223

14. Calculate the volume of these solids, obtained from prisms.

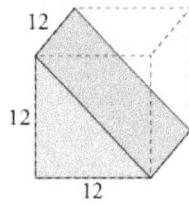

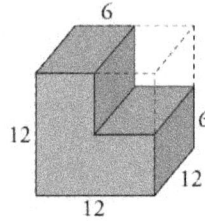

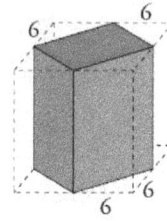

15. Calculate the volume of these solids, whose bases are regular polygons.

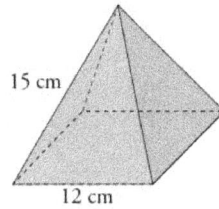

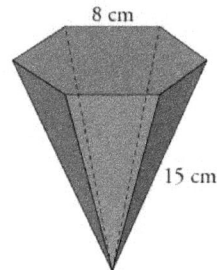

16. Calculate the volume of these solids.

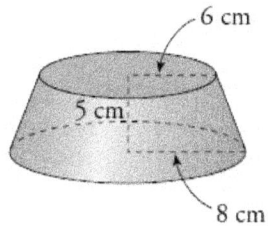

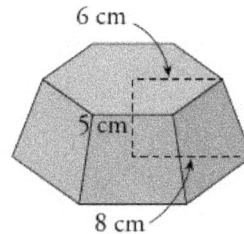

Review exercises

1. Can the figure given below be the development of a cylinder?

2. If we build a cylinder with a rectangular paper sheet, will we obtain the same cylinder by rolling it by its base or by its height?

3. Draw the development of a cone with radius of the base of 5 cm and generatrix of 10 cm.

4. With a 4 cm base and 8 cm height triangle we are generating a cone by rotating it about its height. What is the length of its generatrix?

5. Can the development of a cone be a whole circle?

6. When rotating a quarter of a circle about a radius, what solid will we obtain?

7. When rotating a circle about an exterior axis, what solid are we obtaining?

8. Draw the following solids and calculate their volumes:

a) A cube with edge 9 m. Calculate, also, its area.

b) A regular triangular right prism with basic edge 5 cm and height 16.5 cm. Calculate, also,

   its area.

c) A regular hexagonal right prism with basic edge 8 cm and height 10 cm. Also, its area.

d) A right cylinder with radius 3 cm and height 10 cm.

e) An oblique circular cylinder with radius 3 mm and height 5 mm.

f) A right cone with height 4 cm and base radius 3 cm.

9. Calculate the volume existing between the cube and the cone:

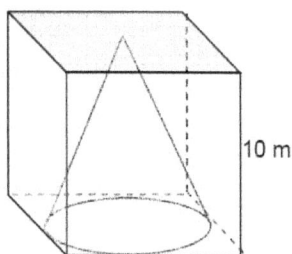

10 m

10. Calculate the area of this figure:

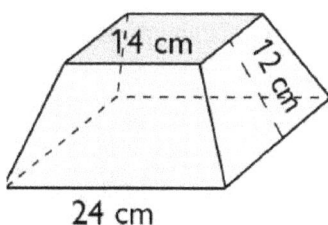

1'4 cm

12 cm

24 cm

225

11. Calculate area and volume of first figure, and volume of second one:

a)

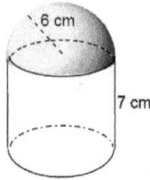

b)

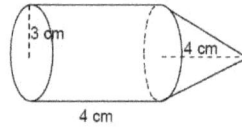

12. a) Estimate the volume and the area of the Earth, knowing its radius is, approximately, 6371 km.

b) Atmosphere has a mean height of 100 km. Estimate its volume and the percent of the total volume that it represents.

13. Calculate volume of Kio Towers, in Madrid, knowing its base is a square of side 35 m and its height is 114 m.

14. We want to paint walls and ceiling of a dining-room whose floor is 12 x 7 m, and whose height is 3.5 m. This room has two doors with dimensions 1 x 2 m, and three windows with dimensions 2 x 2 m.

a) How much area is there to paint? (Draw it).

b) If we have painting cans for 25 m^2, how many cans do we need?

15. Calculate the volume of a Rubik's cube with edge 8 cm.

16. Calculate the volume, in ml, of a can of Coca-cola, whose height is 10.9 cm and whose diameter is 6.2 cm. (1 ml = 1 cm^3)

17. Calculate the volume of Keops' pyramid, knowing its height is 230.35 m and its base is a square of side 136.86 m.

18. A water deposit is a prism whose height is 10 m and whose volume is 4000 m^3. Calculate the side of the base, knowing it is a square.

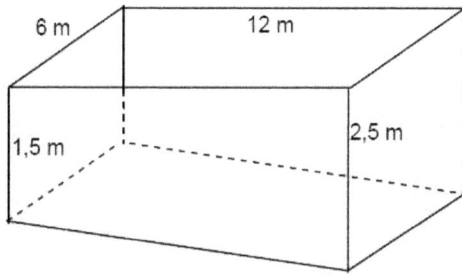

(EXERCISE 19)

19. Calculate the volume of the shown swimming pool.
(1 m^3 = 1 000 l)

20. About previous exercise, how long would it take to fill it if water comes at 0.5 l/s?

21. Base diameter of a cylinder is equal to its height. Total area is 169.56 m^2. Calculate its dimensions.

22. Calculate the area of the following solids:

a)
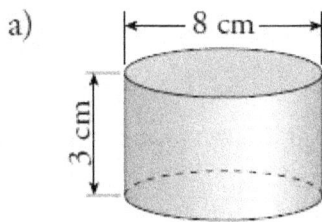

b)
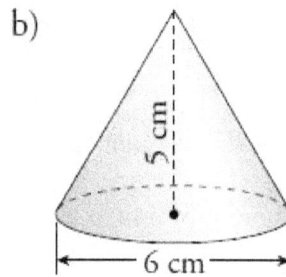

c)
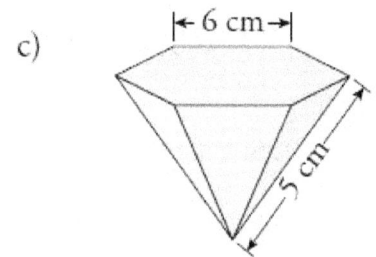

d)
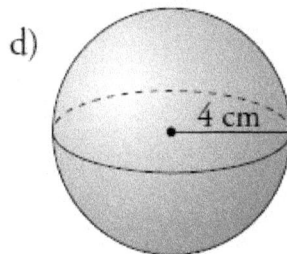

e)
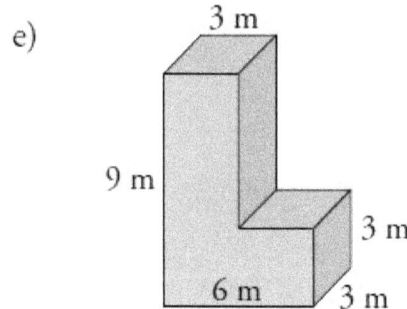

f)
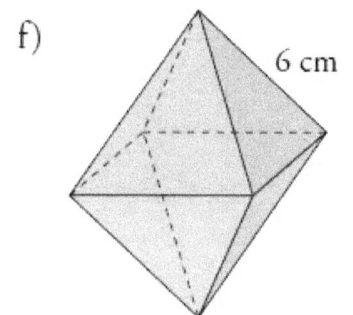

23. Calculate the volume of the following solids:

a)
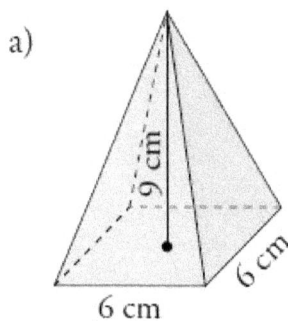

b)
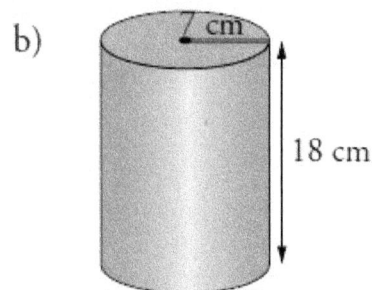

c)
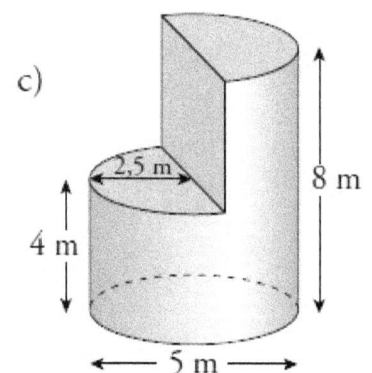

227

d) 8,4 cm 8,4 cm 21 cm 12 cm

e) 12 m 15 m 14 m 16 m

f) 3 m 3 m 3 m 9 m 3 m

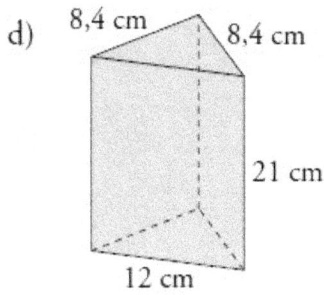

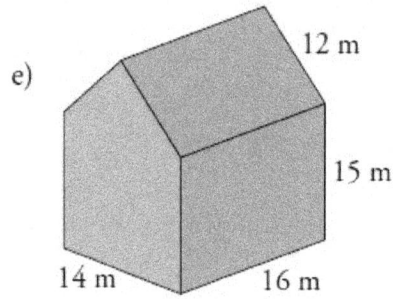

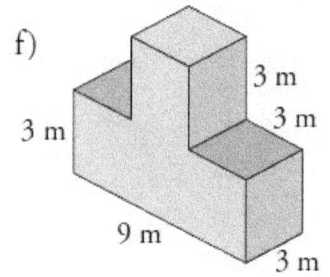

24. Calculate the volume of following solids:

a) Regular octahedron with edge 10 cm.

b) Regular hexagonal pyramid with a lateral edge of 15 cm and basic edge of 8 cm.

c) Semi-sphere with radius 10 cm.

d) Cylinder inscribed in a right prism with a square base with side 6 cm and height 18 cm.

25. Calculate the volume of this regular tetrahedron. Clue: To calculate height H, remember that

$\overline{AO} = \dfrac{2}{3}h$, where h is the height of a face.

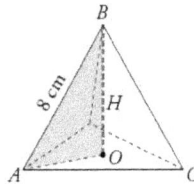

26. Calculate the volume of following solids:

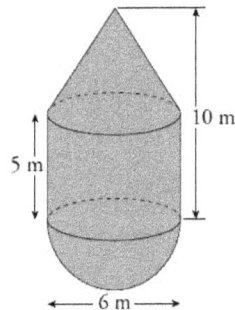

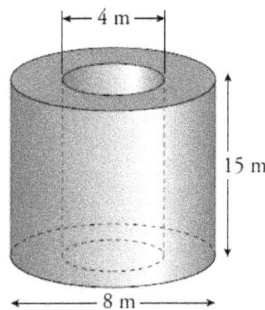

27. Calculate the volume of this solid. Look at the drawing at the right of the solid for help. Units are in meters.

28. Which of these glasses has a higher volume?

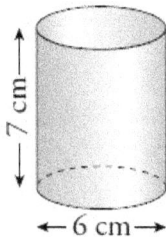

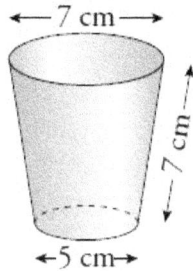

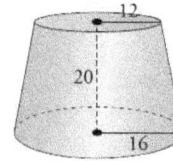

← 7 cm →

7 cm

← 6 cm →

7 cm

← 5 cm →

12

20

16

12

20

16

(EXERCISE 27)

29. What is the area of the coloured triangle?

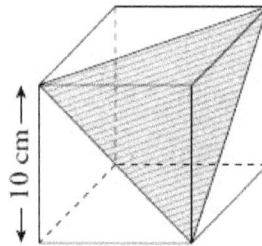

10 cm

30. What is the volume of the room whose floor is shown?

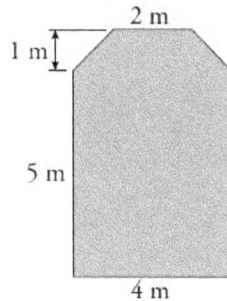

2 m

1 m

5 m

4 m

31. Calculate the volume of the solids generated when rotating these plane shapes around the indicated axis.

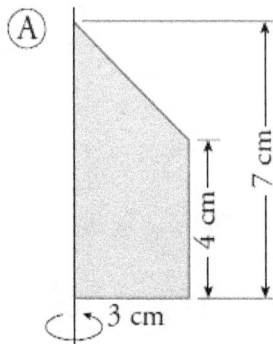

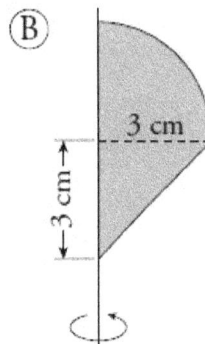

(A)

7 cm

4 cm

3 cm

(B)

3 cm

3 cm

229

32. Three tennis balls are kept in a cylindrical box with 6.6 cm diameter. Calculate the volume of empty space.

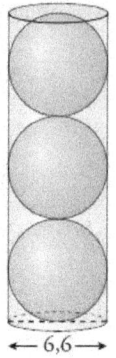

33. We put a 14 cm diameter ball in a 14 cm edge cube, initially full of water. Calculate: a) Volume of dropped water. b) Height of water remaining in the cube.

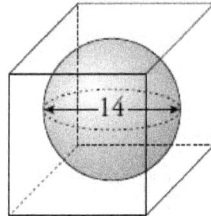

34. A isosceles right triangle has legs of 8 cm. Calcuate the volume of the solid generated when rotating around its hypotenuse.

35. We want to build a cylinder by joining parallel sides of a rectangle whose sides measure 20 and 28 cm. What sides do we have to join if we want to obtain a cylinder with a higher volume?

36. We have cut a 8 cm edge cube as shown in the figure. What is the volume of each portion?

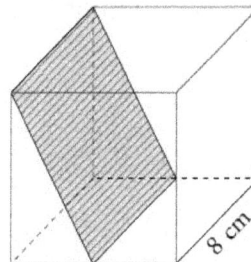

37. When opening a cone we obtain a circle sector with angle 120° and area 84.78 cm^2. Calculate the total area and the volume of the cone.

38. A cylinder and a cone have the same total area, 96π cm^2, and the same radius, 6 cm. Which of them will have a higher volume?

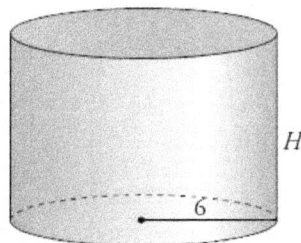

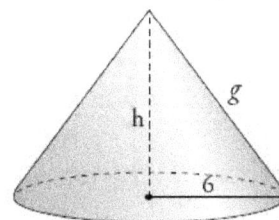

230

39. a) Which of the following solids are polyhedrons?   b) How many of them are regular?

c) Check which of them verify the Euler's formula.

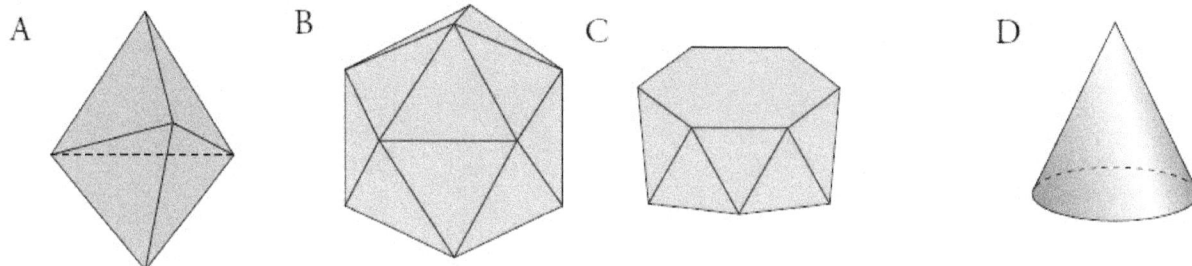

40. If in a cone we reduce into its half the base radius and keep its height, will its volume reduce into its half? And if we keep the base and reduce its height into the half?

41. A square base pyramid is cut by a horizontal plane, parallel to the base and passing by the medium point of the height. What will be relationship between volumes of higher and lower pyramids?

42. A cube and a sphere have the same area. Which of them has a higher volume? Do it by giving a value to the sphere's radius.

43. In a cylinder of diameter identical to the height, we inscribe a sphere. What is the relationship between lateral area of the cylinder and sphere's area?

44. What is the relationship between the volumes of this sphere and this cone?

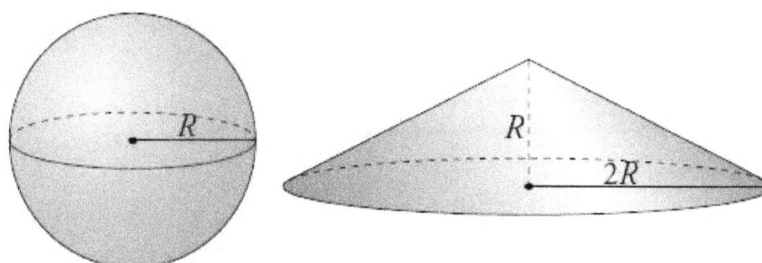

$f(x)=2^x$

$g(x)=\log_2 x$

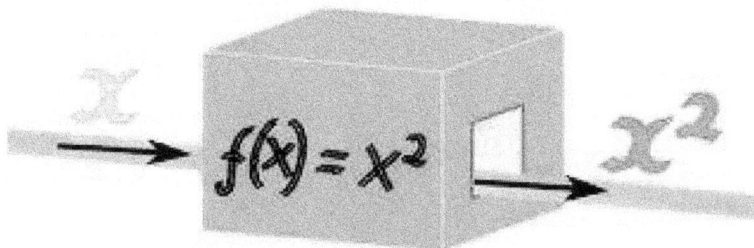

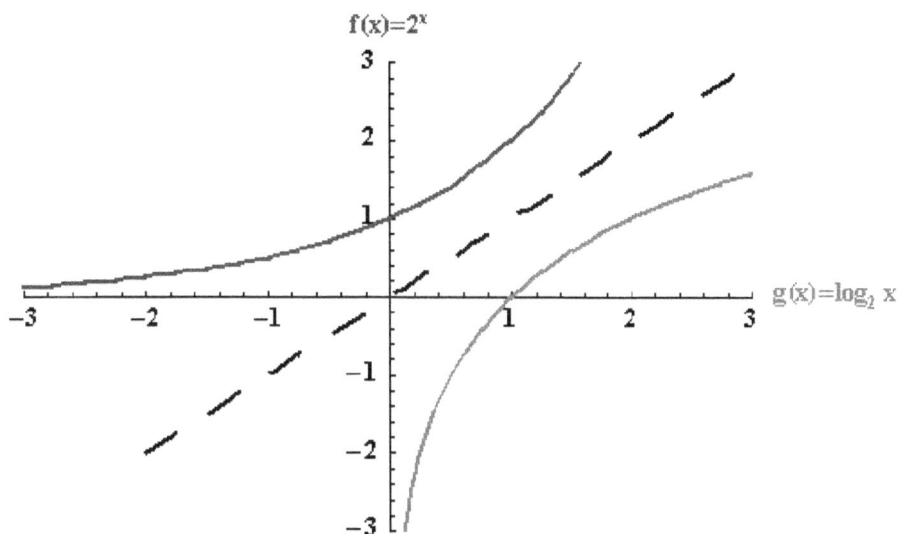

Unit 12.- Functions

1. Starting. Coordinates in a plane plot

Before starting this unit about functions and graphs, we need remember some concepts about them.

Example: Represent the following values in the Real Number Line: 2, 0, -1, 3, -2, 1, -3.

Consider two lines, one of them horizontal and the other one, vertical. They are going to be the reference axes to locate points in the plane $\mathbf{R}^2$. They are named *X-axis* or *abscise axis*, and *Y-axis* or *ordinated axis*.

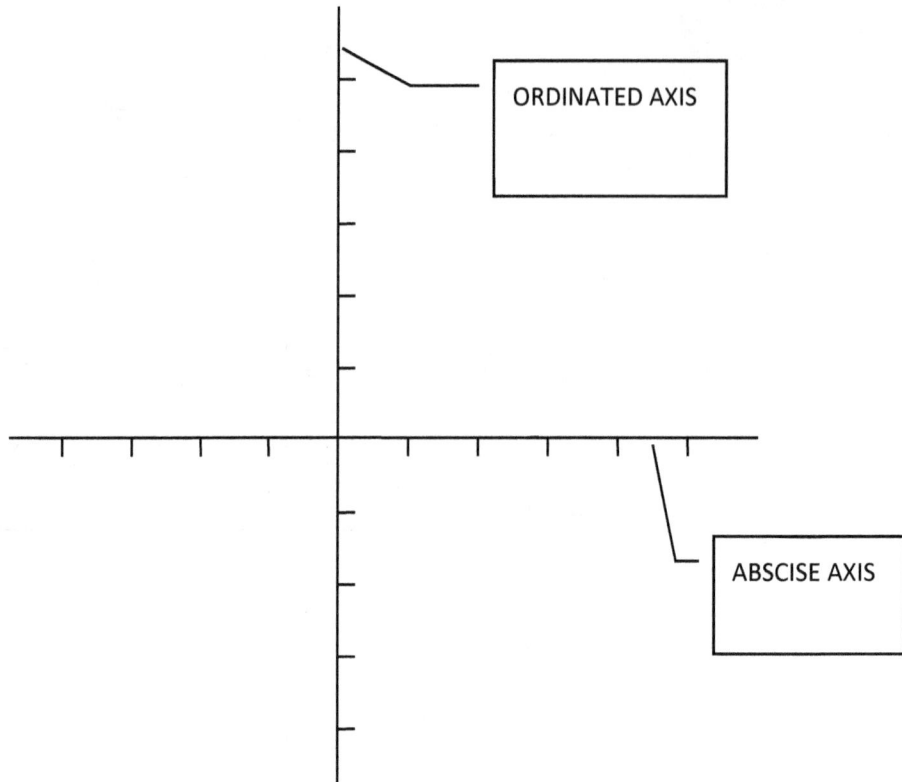

ORDINATED AXIS

ABSCISE AXIS

Points in the real plane, **R**^{2}, are located by denoting them with two numbers or ***coordinates, (x,y).***

Example: Copy the coordinated axes and represent the following points:

A (2,3), B (4,1), C (-3,2), D (-4,-2), E (5,-1), F (3,-3), G (8, -6), H (-7, 5).

Example: A cyclist advances a distance of 70 km in 3 hours and he has been writing down times and distances in a table.

Distance (km)	0	10	20	30	40	50	60	70
Time (min)	0	30	60	90	120	140	160	180

Write down times in the abscise axis and distances in the ordinated axis. You do not need to use the same scale in both axes.

Example: Locate points A to H, and write down coordinates of points I to M.

A(3, 2), B(4, -5), C(-1, 3), D(-7, -2), E(-5, 0), F(4, 0), G(0,-1) y H(0, 1)

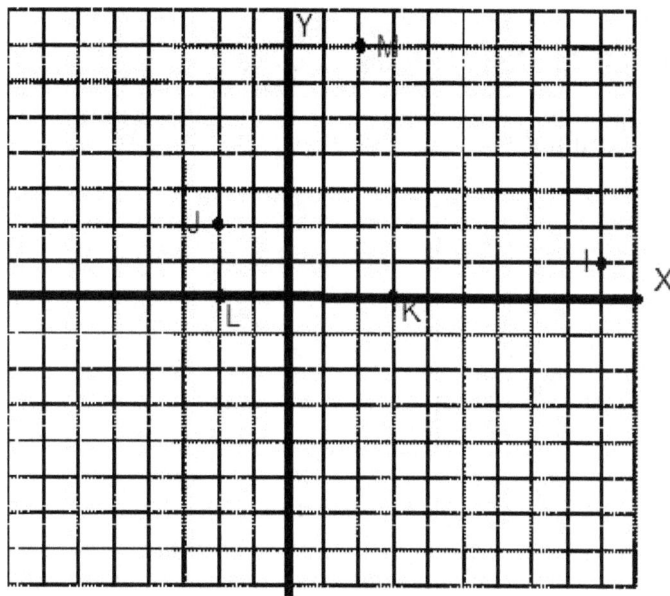

Example: About following graph, write down the images of -4, -3, -2, -1, 0, 1, 2, 3 and 4:

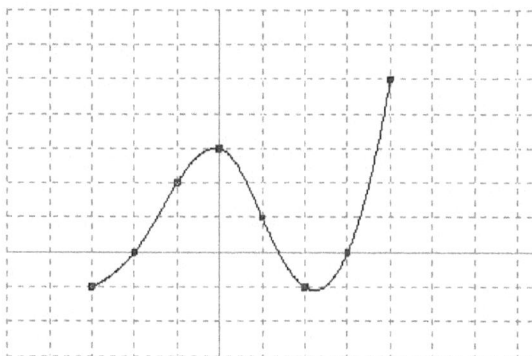

f(-4) =   f(1) =

f(-3) =   f(2) =

f(-2) =   f(3) =

f(-1) =   f(4) =

f(0) =

1. In different coordinated axes systems, draw the following points. Pay attention to scale in each case.

a) $A(3, 2)$, $B(5, 1)$, $C(0, 2)$, $D(5, 5)$, $E(3, 0)$.

b) $A(-3, 5)$, $B(0, -6)$, $C(-1, -3)$, $D(3, 4)$, $E(5, -2)$.

c) $A(3; 0,5)$, $B(2; -2,5)$, $C(-4,5; 2)$, $D(0; 3,5)$, $E(-3,5; -4,5)$.

Exercises

2. Draw the figure obtained when joining each point with its following:

$A(2, 1)$, $B(2, 3)$, $C(3, 3)$, $D(3, 5)$, $E(6, 5)$, $F(6, 3)$, $G(7, 3)$, $H(7, 1)$, $I(5, 1)$, $J(5, 2)$,

$K(4,5; 3)$, $L(4, 2)$, $M(4, 1)$, $A(2, 1)$

3. Write the coordinates of these points:

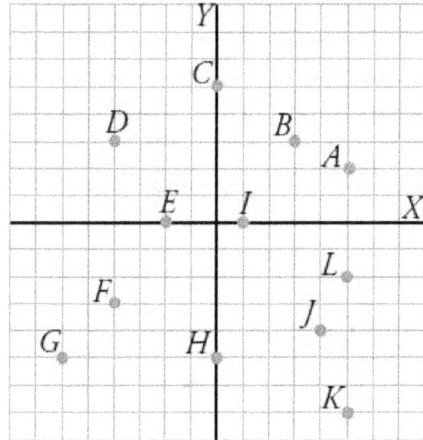

4. Look at points A, B, C and D in this graph:

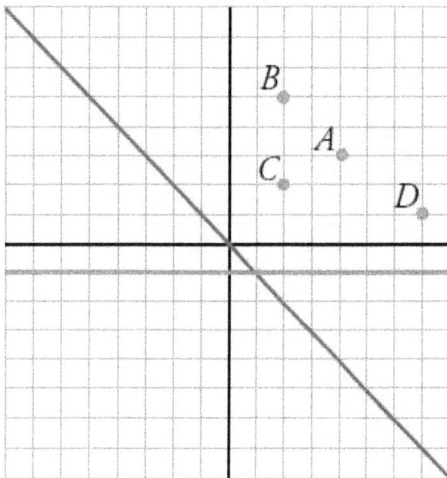

a) Write their coordinates.

b) Draw their symmetric points about the blue horizontal straight line.

c) Draw their symmetric points about the Y axis.

d) Draw their symmetric points about the red inclined straight line.

5. These points represent a cat and an elephant. Indicate the correspondence.

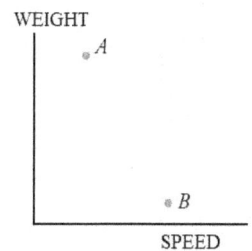

6. A family has just travelled for 4.5 hours. Following graph shows their position depending on time from leaving home.

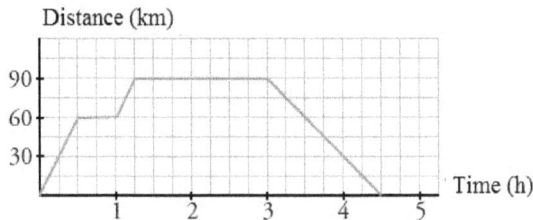

Distance (km)

a) How many km have they done the first half an hour?

b) How long does the car stay stopped?

c) How many km have they done when doing the first stop? And second?

7. Observe this other travel done by another family:

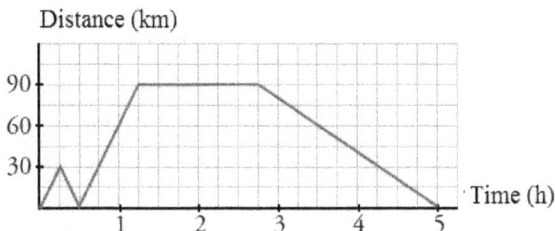

Distance (km)

a) How many km have they done when they decide to come back the first time?

b) Where do they stop? How long does the stop take them?

c) How long time has the car been moving for?

8. For a shoe shop, incomes and expenses are depending on the number of sold pairs. This graph shows the relationship.

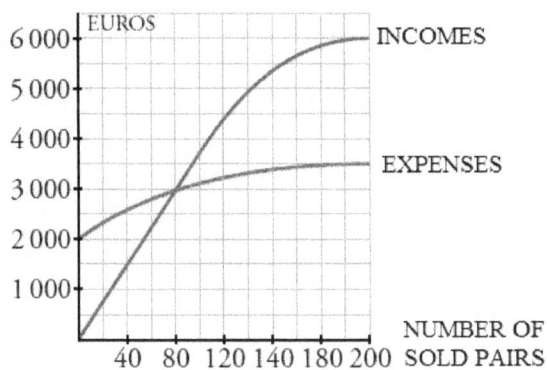

EUROS

INCOMES

EXPENSES

NUMBER OF SOLD PAIRS

a) From how many sold pairs do they begin to have benefits?

b) How much do they lose if they sell only 40 pairs?

c) How much do they earn if they sell 120 pairs?

d) How much do they earn if they sell 200 pairs?

3. Dependence between magnitudes

a) Relationships given by tables

In a laboratory class, a student has measured temperature of a liquid while it was being heated. He wrote results in the following table:

x = time (min)	0	1	2	3	...
y = temperature (°C)	20	24	28	32	...

Temperature of the liquid (*dependent variable*) depends on the time (*independent variable*).

b) Relationships given by a graph

Following graph represents variation of temperature of an ill person in a hospital during a whole day. From it, it is easy to determine the relationship between the time and the temperature of that person. For example, his lowest temperature has been reached at 2 p.m. (14:00 h) and highest one at 8 p.m. (20:00 h).

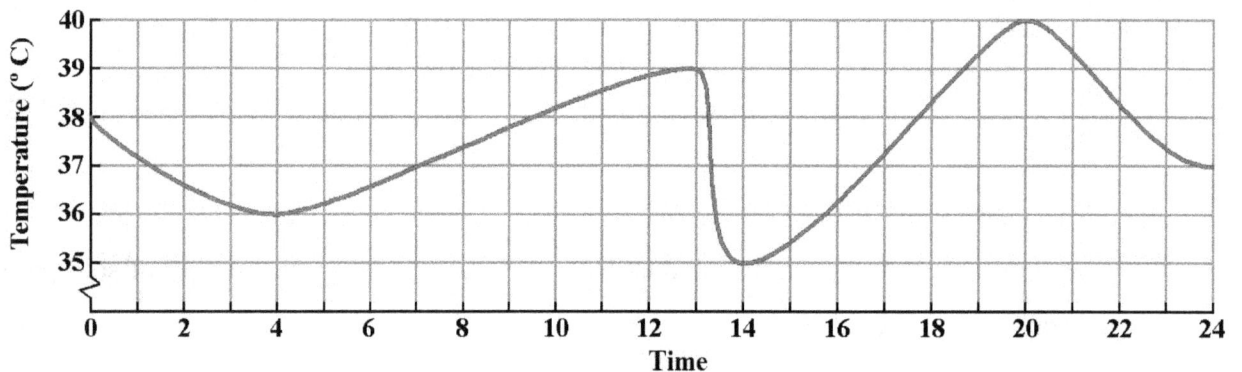

Graph gives us an intuitive view about the relationship between two magnitudes and the behavior of one magnitude, depending on the other one. Again, temperature (*dependent variable*) depends on the time in the day (*independent variable*).

c) Relationships given by formulas

Many of the relationships between magnitudes in Nature are given by algebraic expressions.

- Volume of a sphere depends on its side. The formula that relates both magnitudes is $V = \frac{4}{3}\pi a^3$.

- If price of 1 kg of oranges is 0.35 €, the cost *(C)* of a quantity of oranges depends on the number *(k)* of kilograms. The formula that relates both magnitudes is $C = 0.35k$.

> *Dependent variable* is the one that depend on the other one, named, *independent variable.* Dependent variable is usually denoted with the letter *y*, and independent variable is usually denoted with the letter *x.*

4. Concepts concerning to functions

Above relationships have one common characteristic: each value of the independent variable corresponds an *only* value of the dependent variable. A relationship verifying this characteristic is named *function*.

> A *function* is a relationship or correspondence between two magnitudes, so that each value of the independent variable *x*, corresponds *an only value* of the dependent variable *y*.
>
> In order to indicate that a magnitude *(y)* depends or is a function of another one *(x)*, it is used the notation $y = f(x)$, which is read *"y is a function of x"*.

Functions are like machines in which you introduce an element, *x*, and it gives you back another value, $y = f(x)$. For example, in function $f(x) = x^2$, we introduce *x* values, and it will give us back their squares.

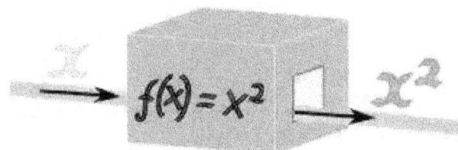

Calculator keys define functions by using formulas. For example, square root key, $\boxed{\sqrt{}}$ defines the function $y = \sqrt{x}$, or $f(x) = \sqrt{x}$.

For example, if we introduce in a calculator 25 and push $\boxed{\sqrt{}}$ we will see in the screen: 5. $\rightarrow$ This means 25 is a valid input and 5 is a valid output.

5 is said to be the *image* or *transformed* of 25 in the function $f(x) = \sqrt{x}$.

It is written *f(25) = 5*.

For example, if we introduce - 4 and push $\boxed{\sqrt{}}$ we will see in the screen: ERROR → This means - 4 is not a valid input for this function.

$y = \sqrt{-4}$ y is not a valid output for this function, there is not an input that gives us back in the screen $\sqrt{-4}$

Example: Given the function *f* that joins each real number to its double:

a) Write its algebraic expression.

b) Calculate the images of 1, 2 and 3, denoted by *f(1), f(2)* and *f(3)*.

Solution:

a) Algebraic expression of this function is denoted by *y = 2x*, or also by $f(x) = 2x$.

b) $f(1) = 2 \times 1 = \boxed{2}$; $f(2) = 2 \times 2 = \boxed{4}$; $f(3) = 2 \times 3 = \boxed{6}$.

5. Graphical representation of functions

Graphs can be given in some different ways:

a) Graphs given by a table

In a book shop, they have the following table with the cost of photocopies, depending on the number of copies:

Number of copies	5	10	15	20	25	30	...
Price (euros)		2.28	3.37	4.00	5.50	6.12	7.00	...

We can build the corresponding plot by representing these points:

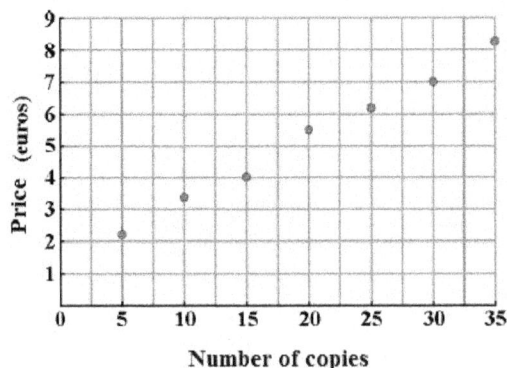

We might have a question: Could I draw intermediate points between these ones? The answer is: not for all, only for a few, because of the nature of magnitudes we are working with. You cannot buy 22.5 copies. You can only by a whole number of copies.

b) Graphs given by a formula

The expression that gives the area of a circle as a function of its radius is $A = \pi \cdot r^2$. It is a function given by a formula. To represent the function, we create a table, freely introduce values for radius and calculate, by using the formula, corresponding areas.

x =radius	...	...	...	...	...	...	...
y = area	...	...	...	...	...	...	...

We represent the obtained pairs of values and get different points of the plot. If we gave intermediate values to the radius, we would obtain intermediate values of area. Remember in previous example, this was not possible, because you cannot buy 1.5 copies. So, in this case, we can join the points, *in a soft way*.

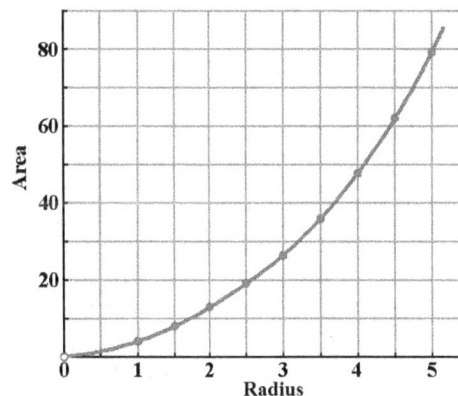

Exercises

9. Explain if following relationships are functions, indicating independent (x) and dependent (y) variables:

a) Each real number is transformed in its cubic root.

b) Each integer number is transformed in its prime factors.

241

10. Which of the following graphs represent a function?

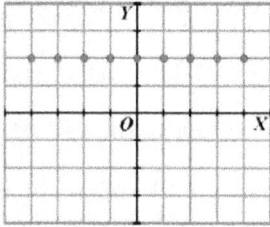

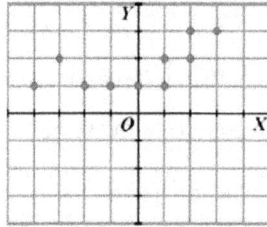

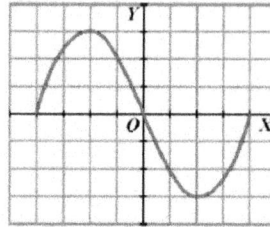

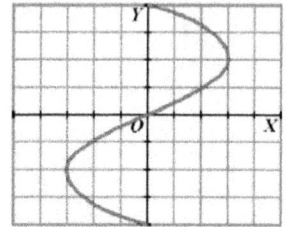

11. Juan has just joined a video-club where he has to pay 6 € as an inscription, 4 € per film (only 10 first) and 2 € per film (for 11th and more).

    a) Write a formula or algebraic expression of the function that relates the number of

       rented films and the cost of them.

    b) Build a table with the cost of renting 0, 1, 2, 5, 10, 20, 50 films.

    c) Represent the plot of this function.

12. The cost of fuel depending on the kilometers length of a car´s trip is given by the following table:

Distance: x (km)	0	50	100	150	200	250	...
Fuel: f (liters)	0	4	8	12	16	20	...

    a) Write the expression of the function $f(x)$,   b) Calculate $f(300), f(400)$ y $f(450)$.

    c) Represent the plot of the function. What kind of plot have you obtained?

13. We want to build a cylinder shaped well with a radius of 2 m. Express the volume of water (y) that the well is able to contain, depending on its depth (x). Represent its plot.

14. Build a table corresponding to all rectangles with area 36 cm^2 whose base and height are integer numbers. Like this one:

Base: x (cm)					...
Height: y (cm)					...

a) Draw the corresponding plot, representing base in x-axis (*abscise axis*) and height in y-axis (*ordinated axis*).   b) Join the points in a *soft* way.

c) By using the previous table, calculate the perimeters of those rectangles. Draw the plot that relates the base of rectangle and its perimeter, representing the base in the abscise axis and perimeter in the ordinated axis.

6. Continuous and discontinuous functions

Remember the plots obtained in the previous epigraph:

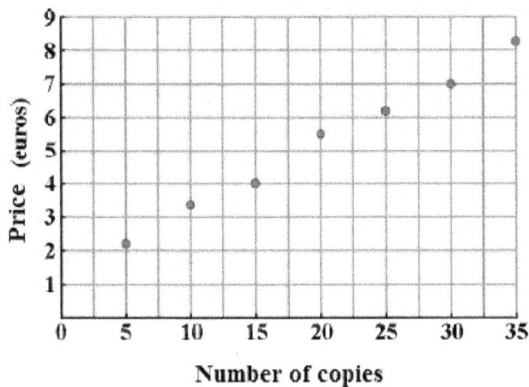

In this first case, independent variable can only take natural numbers and its plot is a series of spots. Functions having this kind of plot are named *discontinuous functions*.

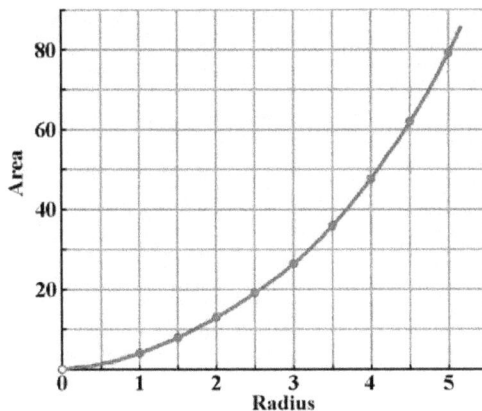

In this other case, independent variable can take all positive real numbers ans its plot is a continuous line. Now we have a *continuous function*.

Imagine now this situation. You are in a telephone box and you are going to phone home. You need 4 coins for the first three minutes. From the third minute, you need a new coin for each three minutes of conversation.

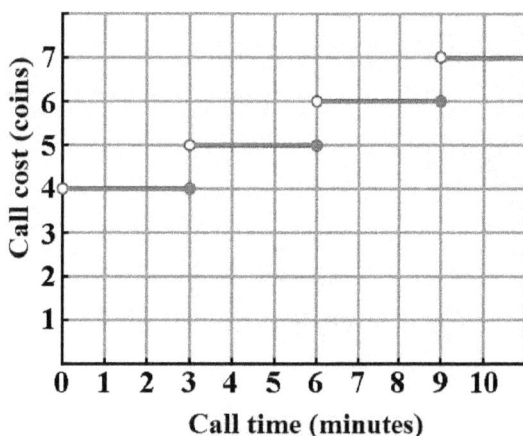

This graph is a discontinuous one. But now, independent variable is continuous (call time), and dependent variable (call cost) shows a series of jumps.

A function $y = f(x)$ is **continuous** if its plot can be drawn as an only continuous line. This means, you can draw its plot without picking your pencil up from the paper.

A function $y = f(x)$ is **discontinuous** if its plot cannot be drawn as an only continuous line and, therefore, it shows *"jumps"* in its graph. Points where are located these *jumps* are named **discontinuity points**.

Exercises

15. Study the continuity of the following graphs:

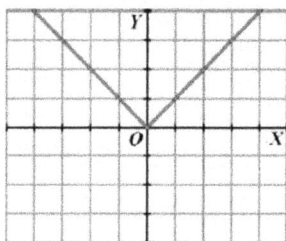

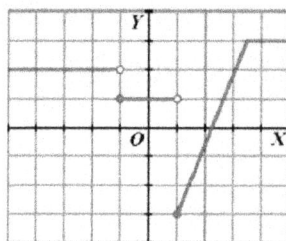

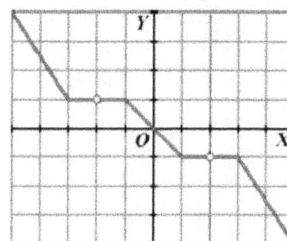

16. Function integer part y = E(x) is defined as the function that relates a real number to its immediately lower integer number. So, for example:

E(0.6) = 0; E(2.8) = 2; E(-5.7) = - 6; E(5) = 5.

Graphical representation of this function is shown on the right. Study its continuity.

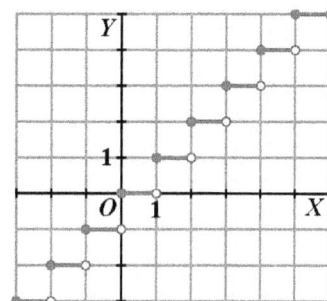

244

17. The cost of sending a telegram is 1.20 € as a fix tax plus 0.05 € per word. Build a values table with the cost of sending a telegram, depending on the number of words. Is it a continuous function? Why?

18. In a book shop they have different prices for copies of an only original. First 100 have a price of 5 cents/copy, from 101 to 200 copies, 4 cents, from 201 to 500, 3 cents, and more than 500, 2 cents. Represent the function that relates the number of copies with the price per unit.

---

7. Variations of a function

Introduction: intervals and semi-straight lines

Intervals are segments into the Real Line. There are different types of intervals:

The open interval **(a,b)** includes the real numbers x between a and b, but a and b are NOT included. It is expressed as $a < x < b$.

The closed interval **[a,b]** includes the real numbers x between a and b, being a and b included. It is expressed as $a \leq x \leq b$.

In the same way, the interval **[a,b)** is expressed as $a \leq x < b$, being a included but not b;

while the interval **(a,b]** is expressed as $a < x \leq b$, being b included but not a.

The open semi-straight line **(a, ∞)** includes all the real numbers being higher than a, where a is not included. It is expressed as a < x;

a

while the closed semi-straight line **[a, ∞)** includes all the real numbers being higher than a, being a included. It is expressed as a ≤ x.

a

In the same way, semi-straight line **(-∞, b)** includes all the real numbers being lower than b, where b is not included. It is expressed as x < b;

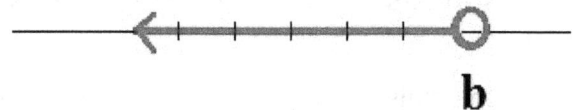

b

while the semi-straight line **[-∞, b]** includes all the real numbers being lower than b, being b included. It is expressed as x ≤ b.

b

Increasing and decreasing. Extremes. Maximums and minimums

A function $y = f(x)$ has a ***maximum*** in a point $x = x_0$ if, in values near it, to the left of x_0, function is increasing, and to the right of x_0, function is decreasing. This means that in points near it, function takes values always lower than it.

A function $y = f(x)$ has a ***minimum*** in a point $x = x_0$ if, in values near it, to the left of x_0, function is decreasing, and to the right of x_0, function is increasing. This means that in points near it, function takes values always higher than it.

You are going to see it more clearly in the following figures.

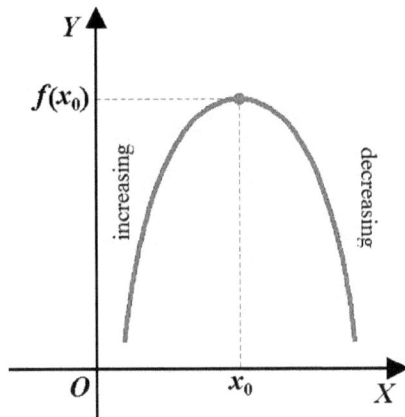

Maximum in x = x_0

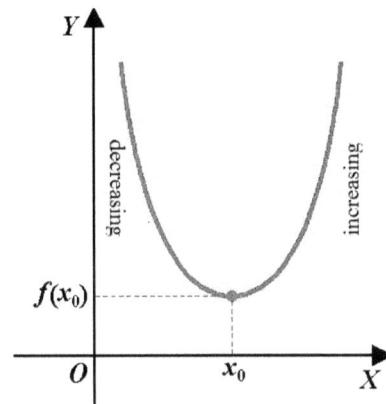

Minimum in x = x_0

Anyway, a function can have several minimums and maximums. In order to distinguish them, we define the following concepts.

· A function $y = f(x)$ has an ***absolute maximum (minimum)*** in a point $x = x_0$ if <u>all</u> the values that the function takes are lower (higher) than its image $f(x_0)$.

· A function $y = f(x)$ has a ***relative maximum (minimum)*** in a point $x = x_0$ if <u>near to it</u>, values that the function takes are lower (higher) than its image $f(x_0)$.

Exercises

19. Next graph shows the profile of a day at the *"Tour de France"*. Study the increasing and decreasing of the function and its maximums and minimums.

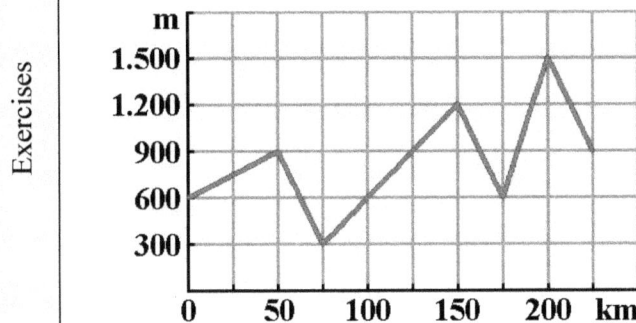

8. Intersection points with the axes

When studying functions, it is interesting knowing if they intersect axes and where they do it.

· Intersection points with Y-axis

Intersection point of a function $y = f(x)$ with the Y-axis has abscise = 0. So, its coordinates are $(0, f(0))$.

· Intersection points with X-axis

Intersection point of a function $y = f(x)$ with the X-axis has ordinated 0. So, they are the points whose x coordinate are solution of equation $f(x) = 0$.

Example: The following graph intersects the coordinated axes at points Q(0, -3) and P(5, 0).

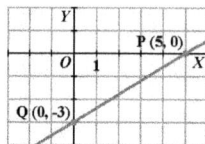

---

21. Study the plot of the following functions, indicating:

a) Continuity.   b) Increasing and decreasing intervals.

c) Maximums and minimums (relative and absolute ones), indicating the value of the

   function in those points.

d) Intersection points with the axes.

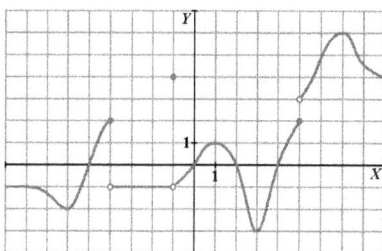

9. Linear function y = mx

AVE train advances with an average speed of 240 km/h. The following table shows advanced distance as a function of trip time.

t = time (h)	1	2	3	4	5	...
e = distance (km)	240	480	720	960	1200	...

As you can see, these magnitudes are in a direct proportion. The algebraic expression of this function is $e = 240t$, where 240 is the proportionality constant. Plot of this function is shown below:

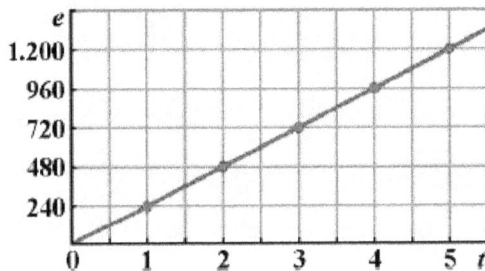

Graph of this function is a continuous *linear* plot. This concrete case is an increasing linear plot (distance increases when time increases) and it passes by the origin of the coordinated axes.

Direct proportionality functions or *linear functions* are those ones whose graph is a linear plot that passes by the origin of the coordinated axis (0, 0).

Its algebraic expression is $y = mx$, where *m* is the proportionality constant.

9.2. Slope of a linear plot

In a linear function $y = mx$, coefficient *m* (proportionality constant) is named *slope* and it can be calculated by dividing values of dependent variable by their corresponding values of independent variable.

$$m = \frac{y}{x}$$

Its value indicates the rhythm of increasing or decreasing of the linear plot with equation $y = mx$. It indicates how *y variable* changes when *x variable* changes.

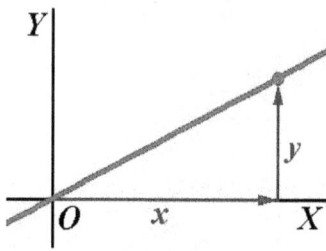

$m > 0 \rightarrow$ linear plot is **increasing** $m < 0 \rightarrow$ linear plot is **decreasing**

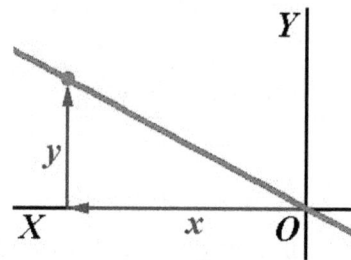

Slope of a linear plot is related to its inclination about the X-axis (angle formed with this axis).

In the following example, you can see that the higher is the slope the higher the inclination of the plot is.

Example: Work out the algebraic expression of the linear functions whose graphical representation is given below:

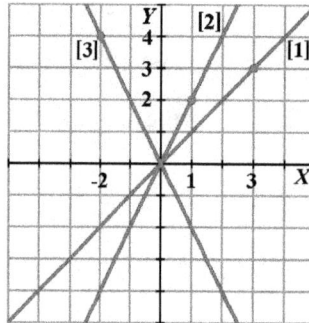

Solution:

Three plots are linear functions with equation $y = mx$, because they are linear plots passing by the origin (0, 0).

Slopes are calculated by working on a point of the plot and calculating the quotient:

(1) $m = \dfrac{3}{3} = 1$;   (2) $m = \dfrac{4}{2} = 2$;   (1) $m = \dfrac{4}{-2} = -2$;

Linear plots have equations: (1) y = x; (2) y = 2x; (3) y = -2x.

22. In a shop, 21 meters of a wire are sold by 3 €.

a) How much is each meter of wire?

b) Build a table indicating the cost of 1, 2, 3, 4, 5, … meters.

c) Draw the corresponding plot and check it is a linear function.

d) Write the algebraic expression of this function. What is the slope or proportionality

constant?

23. Following table shows cost and number of photocopies made in a book shop by some

students.

x = number of copies	20	40	60	80	…
y = price (euros)	0.80	1.60	2.40	3.20	…

Find the expression that relates the number of copies and their cost. Draw it.

24. Following plots show the relationship existing between volume and mass for some

materials.

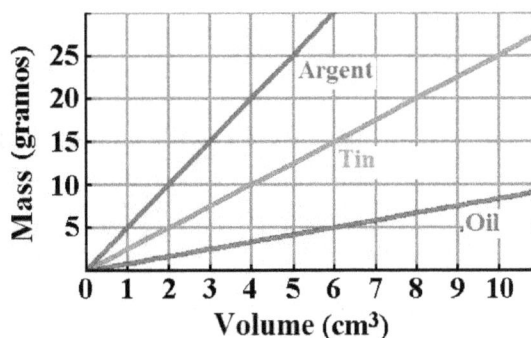

a) Calculate slopes of each linear plot. What is the most dense material? And the least

dense? Write down the algebraic expression for all of them.

b) What is the weight of 3 dm³ of argent?

c) How many liters are occupied by 1 kg of oil?

25. Determine the expression for the functions whose plots are given below.

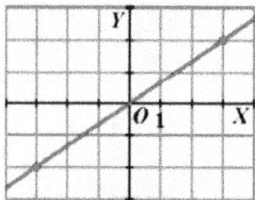

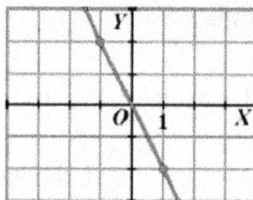

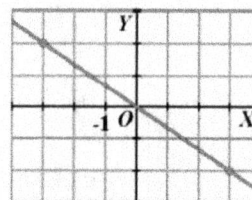

26. Write an equation for each of the following linear functions and draw them on an only coordinated axes:

a) Linear function passing by origin and its slope is 1/2.

b) Linear function passing by origin and by point (-1, 3).

c) Linear function symmetric to $y = 2x$ about the ordinated axis.

d) Linear function symmetric to $y = 2x$ about the abscise axis.

10. Affine function y = mx + n

We have measured temperature of a liquid while it was been heated. Results are shown in the following value table.

x = time (min)	0	1	2	3	4	...
y = temperatura (ºC)	15	20	25	30	35	...

Algebraic expression for this function is $y = 5x + 15$ whose graphical representation is shown below.

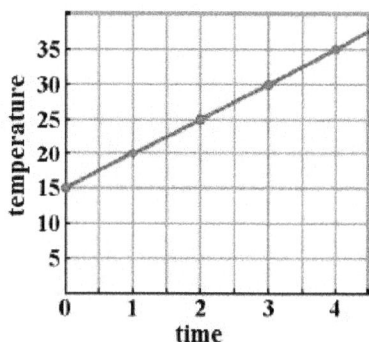

This is not a linear function. Its graph is also a straight line but it does not pass by the coordinate's origin. Its algebraic expression is not $y = mx$, but $y = mx + n$. These functions are named *affine functions.*

Affine functions are those ones whose graph is a straight line that does not pass by the coordinate's origin. Its algebraic expression is $y = mx + n$.

In above expression:

· m is the slope or gradient of the function.

· n is the y-intercept: straight line intersects ordinated axis at point $(0, n)$.

Equations of affine functions

Calculating the slope from 2 points

Given two points with coordinates $A(x_1, y_1)$ and $B(x_2, y_2)$, we can calculate the slope of the straight line that joins them. We have only to calculate variation of y-variable (increasing or decreasing) when

x- variable varies, passing from point A to B.

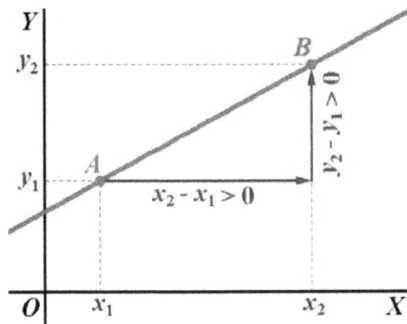

Slope, m, of a straight line passing by points $A(x_1, y_1)$ and $B(x_2, y_2)$ is calculated as

$$m = \frac{y_2 - y_1}{x_2 - x_1}$$

Example: Calculate the slope of the straight line passing by points with coordinates $A(-1, 1)$ and $B(1, 5)$.

Solution:

$$m = \frac{y_2 - y_1}{x_2 - x_1} = \frac{5-1}{1-(-1)} = \frac{4}{2} = 2$$

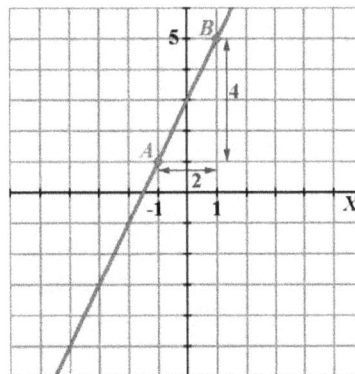

253

Equation of the affine function from 2 points

There are two methods to calculate the equation for the straight line passing by 2 points.

Example: Write the equation for the straight line passing by points with coordinates $A(-1, 1)$ and $B(1, 5)$.

Method 1:

We calculate the slope: $m = \dfrac{y_2 - y_1}{x_2 - x_1} = \dfrac{5-1}{1-(-1)} = \dfrac{4}{2} = 2$

Affine function equation is: $y = mx + n$.   So, $y = 2x + n$.   We only need to know n value.

We freely choose one of the points by which this function passes. For example: $B(1, 5)$.

As function passes by B, its coordinates need to verify equation of this function. We have only

to substitute: $y = 2x + n \rightarrow \begin{vmatrix} x = 1 \\ y = 5 \end{vmatrix} \rightarrow 5 = 2 \cdot 1 + n \rightarrow 5 = 2 + n \rightarrow n = 3$

So our equation is $\boxed{y = 2x + 3}$.

Method 2:

As function is passing by points A and B, their coordinates need to verify equation $y = mx + n$
If we substitute coordinates of points A and B in this equation, we will have a system of linear equations.

$A(-1,1)$: $y = mx + n \rightarrow \begin{vmatrix} x = -1 \\ y = 1 \end{vmatrix} \rightarrow 1 = m \cdot (-1) + n \rightarrow 1 = -m + n \rightarrow -m + n = 1$

$B(1,5)$: $y = mx + n \rightarrow \begin{vmatrix} x = 1 \\ y = 5 \end{vmatrix} \rightarrow 5 = m \cdot 1 + n \rightarrow 5 = m + n \rightarrow m + n = 5$

We have the following system: $\begin{cases} -m+n=1 \\ m+n=5 \end{cases}$ Its solution is ($m = 2$, $n = 3$).

So our equation is $\boxed{y = 2x + 3}$.

Equation of the affine function from its slope and y-intercept

Example: Find the equation of the straight line being parallel to the one with equation $y = 4x - 2$, knowing it passes by the point with coordinates (1, 9).

Solution:

The straight line we are looking for is parallel to $y = 4x - 2$. So, its slope is $m = 4$, and its equation will be $y = 4x + n$.

As it passes by point (1, 9), this coordinates must verify $9 = 4 \cdot 1 + n$, so $n = 9 - 4 = 5$.

Our straight line is $\boxed{y = 4x + 5}$.

---

27. Find the equations of the following straight lines and draw them on an only coordinated axes:

a) Straight line whose slope is 3 and y-intercept is 2.

b) Straight line whose slope is 2 and passes by point (2, 7).

c) Straight line being parallel to straight line with equation $y = 3x - 5$ and passing by the point (-2, 3).

d) Straight line passing by the point (3, 0) and its y-intercept is 3.

e) Straight line passing by the points with coordinates $A(-3, 5)$ y $B(1, 5)$.

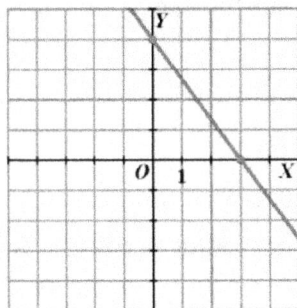

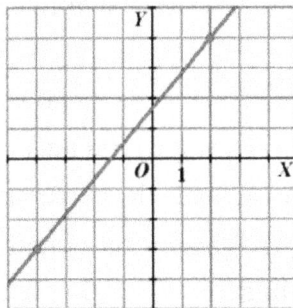

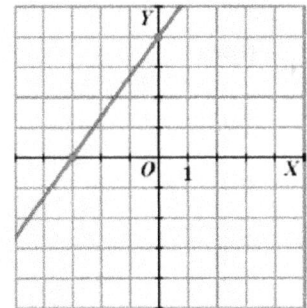

28. Determine expressions for functions whose plots are given below:

29. Given straight line $y = 2x + 4$, find equations for the following straight lines (it might be useful a graphical scheme)

    a) Symmetric straight line to the given one, about *Y-axis*.

    b) Symmetric straight line to the given one, about *X-axis*.

    c) Symmetric straight line to the given one, about *coordinate's origin*.

30. Decide if the relationships between the following pairs of magnitudes can be represented by a linear of affine function. Write the expression of the function.

a) The total weigh of a delivery of 20 kg potatoes bags and the number of bags in it.

b) The initial payment in a gym is 10 euros. Each month, you must pay 30 euros. The total amount paid as a function of the number of months you have been a member.

c) A meter of a certain rope costs 0.5 euros. The cost of a rope and its length.

d) Monthly, I pay to my mobile company a fix quantity of 5 euros and each conversation minute costs 12 cents. The relationship between the number of conversation minutes and the amount I monthly pay.

Review exercises

1. Lung capacity measures the volume of air we can expulse form our lungs, and it is measured with an instrument named spirometer. Spirometer registers volume of air that goes into lungs and later goes out of lungs. Graph gives us the volume (liters) into the lungs as a function of time (min).

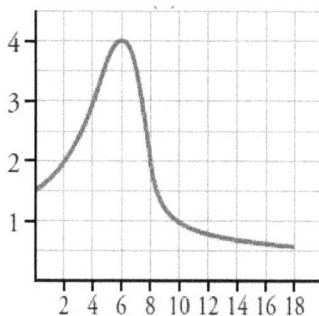

a) What is the initial volume?
b) How long did the observation take?
c) What is the maximum lung capacity of this person?
d) What is its volume 10 minutes after starting? When does it finish?

2. Luis takes 2 hours in going from home to a city 200 km far. There, he has had a business meeting, so, he has stayed at the city for 2 hours. After that, he has gone back home, what has taken him 4 hours.

a) Represent the graph *time-distance to home.*

b) If his speed has been constant during the movement home-city, what is that speed?

c) And in city-home movement?

3. In a city, the want to promote the use of bicycles, so, you can rent a bike. These are prices:

First 3 hours	Free
From third hour	1 €/h

You can rent a bike for a maximum time of 12 hours a day. Draw the plot of the function *time-cost.*

4. Which of the following plots correspond to a function?

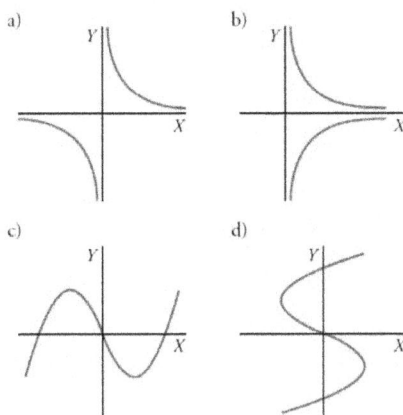

5. For each of the following two graphs, answer to the following questions:

a) Increasing and decreasing intervals.

b) Do they have maximum, minimum or both of them?

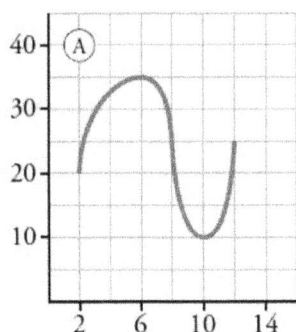

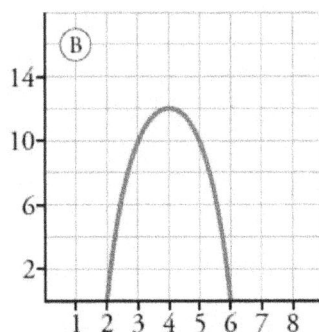

6. Find the slope and write the equations for these straight lines:

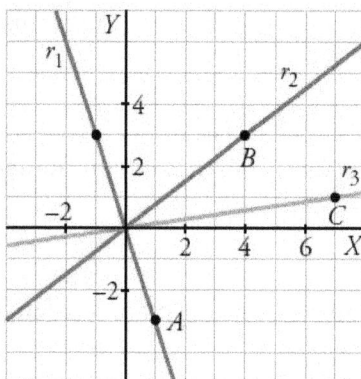

7. Write the equation for the straight line passing by the origin and by the point:

    a) $P\,(12, -3)$ b) $P\,(-2, 3/4)$ c) $P\,(-7, -21)$ d) $P\,(30, 63)$

8. Write the equation of the represented functions. Which are increasing and which decreasing? Think on the sign of their slope.

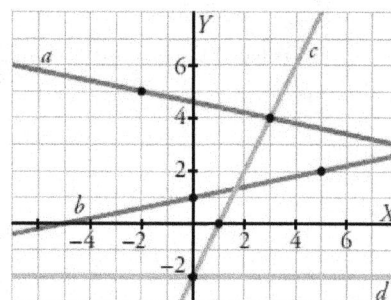

9. Write the equation of all these straight lines and draw them:

    a) Passing by $(-3, 2)$ and $(1, -4)$.   b) Passing by $(2/5, -1)$ and slope $-1/2$.

    c) Passing by $(2, 1)$ and its y-intercept is 3.   d) Passing by $(2, -4)$ and being parallel to $y = 3x$.

    e) Parallel to X-axis and passing by $(-2, -4)$. f) Parallel to Y-axis and passing by $(-2, -4)$.

10. In an English school you must pay an initial quantity of 10 € plus 15 €/month.

    a) Find the expression of the function *number of months* → *Total cost*.

    b) Draw it.

11. In a swimming pool there are 200 m^3 of water. When opening the tap, it empties with a constant rhythm of 40 m^3/min.

a) How long does it take to get empty?

b) Write the algebraic expression of the function that relates quantity of water in the swimming pool, with time.